CITY AND COUNTRYSIDE

Selected, peer reviewed papers from the
2nd International Conference on
Urban Planning and Sustainable Development
(**UPSD** 2013)
May 15, 2013, **Harbin**, **China**
Nov. 9, 2013, **Alghero**, **Italy**

Edited by
Alessandra Lai   Ming Lu

图书在版编目(CIP)数据

城市与乡村=City and countryside:英文/(意)赖(Lai,A.),陆明主编. —哈尔滨:哈尔滨工业大学出版社,2015.2

ISBN 978-7-5603-5069-1

Ⅰ.①城… Ⅱ.①赖… ②陆… Ⅲ.①城乡规划-可持续性发展-国际学术会议-文集-英文
Ⅳ.①TU98-53

中国版本图书馆 CIP 数据核字(2014)第 292851 号

责任编辑 王桂芝 任莹莹
出版发行 哈尔滨工业大学出版社
社 址 哈尔滨市南岗区复华四道街 10 号 邮编 150006
传 真 0451-86414749
网 址 http://hitpress.hit.edu.cn
印 刷 哈尔滨市工大节能印刷厂
开 本 787mm×1092mm 1/16 印张 16 字数 370 千字
版 次 2015 年 2 月第 1 版 2015 年 2 月第 1 次印刷
书 号 ISBN 978-7-5603-5069-1
定 价 78.00 元

Sponsored by

Autonomous Region of Sardinia
Embassy of Italy in China
University of Sassari
Harbin Institute of Technology

Conference Organization

Co-chairman

Prof. Hong Leng	Harbin Institute of Technology, China
Prof. Arnaldo Cecchini	University of Sassari, Italy

Conference Scientific Committee

Prof. Plinio Innocenzi	Science and Technology Councellor, Embassy of Italy
Prof. Alessandro Plaisant	University of Sassari, Italy
Prof. Arnaldo Cecchini	University of Sassari, Italy
Prof. Roberto Pagani	Politecnico of Torino, Italy
Prof. Hong Leng	Harbin Institute of Technology, China
Prof. Huiyi Tang	Harbin Institute of Technology, China
Prof. Ming Lu	Harbin Institute of Technology, China
Prof. Wen Cheng	Harbin Institute of Technology, China
Prof. Kangli Wu	Harbin Institute of Technology, China

Preface

The master's degree program "Advanced Methods and Tools for Sustainable Planning": an educational research laboratory on sustainability at the service of the territory.

In 2010, the definition of the Memorandum of Understanding among the Italian Ministry of Education, University and Research (MIUR), the Autonomous Region of Sardinia (ARS) and the Sardinian Universities for the implementation and the enhancement of the university system and research in Sardinia, offered the opportunity to encourage dialogue and cooperation among universities, institutions and enterprises, by putting in place tools addressed to strengthen the system of research and scientific and technological innovation in strategic sectors for the regional economy, according to the recommendations of the Lisbon Strategy, the National Research Program, the EU's Seventh Framework Programme for Research (FP7). The year after, the Memorandum of Understanding for Higher Education was signed among the ARS Department of Employment, Vocational training, Cooperation and Social Security and the Sardinian Universities to design an innovative project with the aim to increase the postgraduate training offer, focusing on the enhancement of skills and expertise in strategic sectors, with a view of integration in the economic and cultural European and international context.

November 2012 saw the start of the 2nd level University master's degree program "Advanced tools for sustainable planning", designed by the Department of Architecture, Design and Urban Planning of Alghero (DADU) and carried out together with the academic Chinese partner Harbin Institute of Technology (HIT), with the support of the Department of Employment, Vocational training, Cooperation and Social Security of the RAS. It represents an important opportunity for the exchange of views about planning methodologies, current regulations and, above all, the different modalities for a sustainable design and the future management of the city, the environment and the landscape. The idea which inspired the master's degree program is to offer a critical reading of initiatives and processes of implementation, at international level, in order to deal with issues related to different aspects of sustainable planning and management of the landscape in Sardinia and to raise awareness of the current critical situation. In this sense, the Master program offers multiple and interdisciplinary interpretations and points of view, both the technical issues and the opportunities and risks in relation to the strategic dimension of the policies, for example, those which deal, in particular, with the definition of sustainable systems adaptable to different territorial needs, the cooperation between researchers and experts and the training, including the technical structures of local administrations.

Today we cannot think of dealing with issue like the sustainable development without the cooperation with emerging Countries through the exchange of knowledge, appropriate experience and technology to study strategies for future oriented to improve the quality of life, without distinction between urban metropolis and rural village. Beyond the old friendship of our institutions, the implementation of exemplary interventions at different scales, on the one hand, and the cultural and environmental distance of China from Sardinia, on the other hand, leaded us to choose the city of Harbin, in the Heilongjiang Province, being the seat to one of the most prestigious Chinese academic institutions, as the destination for the internship required by the training Master's program. This experience was not confined to the mere knowledge of theories and relevant case studies in Chinese territory, but it served as an input device for the construction of a new vision of the territories, with particular attention to the relationships that govern the urban and the rural dimension of the landscapes. Moreover, the great vitality of this nation with respect to the design of its future shows itself not only in the fast growth of the recent years, along with rapid and simultaneous upheavals both of the society and the environment, which has made China the second largest economy in the world, but also in the care the Country puts today on the issue of "green" economy: in fact, it invests more, in absolute and percentage terms, for the development of environmentally sustainable technologies.

However, this opportunity meets the necessary requirement to build several instrumental in-depth operations involving technological and normative innovation, the reform of the operating modes of public administrations and the effectiveness of protection compared to the models of development, as well as qualified skills related to new ways of understanding the sustainable planning, like the whole project strives for. Specifically it was arranged as an opportunity to rethink the specific field of higher education, meeting the most heartfelt demand to compose shared in-depth analyses, that give a new sense to the instrumental action of the learners in the training project, specialists in architectural and engineering field in particular, but including also specialists in of other fields (agronomists, political scientists, naturalists, etc.), all engaged as members of interdisciplinary teams to deal with procedures and tools oriented to sustainability, declined according to different points of view (not only environmental, but also social, institutional, and, ultimately, economic sustainability) and according to the different perspectives offered by the comparison between Europe and China. These perspectives direct the framework of the future strategic planning and allow to select tools and procedures needed to innovate the design and the sustainable management of landscapes.

Nevertheless the educational project that guides all the activities aims to up-grade the skills of the participants, to the highest degree allowed by state of the art, with regard to the design and the sustainable management of territorial transformations. These have as a natural consequence the landscapes, both in terms of conservation and in terms of transformation. Furthermore traditional models of education, including those who present themselves with a qualifying image, often prove to be unsatisfactory for the learners, due to the lack of an educational reference model. The risk is a contemplative attitude of learners with reference to the knowledge transmitted almost exclusively through conventional forms of training and often in unsuitable spaces to host innovative ways of learning.

Today technical and conceptual structures have not to be trained with an activity made only by frontal lectures, but rather through real designing and planning workshops, which create environments of mutual learning and study, addressed to develop together awareness and testing forms through a common project, without which any interaction is possible. Consequently, if we want to accelerate the EU's transition towards an economy and a society based on knowledge, crucial goal of the Lisbon strategy, the education and training should aim at the construction of interactive environments to build, to maintain, to renew the human and social capital and to promote research excellence in the field of sustainability.

Therefore, the main task of the master is to support the construction of a permanent space of interaction among the people who operate and work in the field of sustainability at different levels and at different scales: institutions, experts, territorial entities and subjects which deal with the territorial management. Specifically, the goal is to put high value and allow researchers and local and international experts to interact, through an efficient network, exchanging their views on experiences and common problems and strategies to cope with them, starting the discussion on the development of new sustainable systems, adaptable to different local needs, identifying lines of action and shared rules to rethink the ways and tools for designing and planning.

Finally, the definition of these tools and methodologies and their possible application is the topic of the Conference "Urban Planning and Sustainable Development" (UPSD

2013). It is an important international meeting point within the master course organized to consolidate a stable and fruitful relationship of scientific collaboration between the two institutions—the University of Sassari and the Harbin Institute of Technology, with a leading role played by the Embassy of Italy in Beijing.

Alessandro Plaisant
Department of Architecture, Planning & Design-University of Sassari (Italy)

Preface

The contributions presented in this book deepen the reflections exchanged during the two conferences organized as part of the International Master's Degree "Advanced methods and tools for sustainable planning", which has consolidated, as just said, the academic partnership between the Department of Architecture, Design and Urban Planning of the University of Sassari (Italy) and the School of Architecture Institute of Technology, Harbin (China).

The first conference was hosted in Harbin May 15, 2013 and the second in Alghero November 11 of the same year.

The two initiatives are also placed in the round of Sino-Italian biennial meetings dedicated to sustainable urban planning (UPSD) and have been encouraged and supported by the Autonomous Region of Sardinia and the Italian Embassy in Beijing in the belief that, despite the profound differences, the constant review of the research and the methodological approach, through the sharing of the results obtained, represents a necessary step for those who pursuing excellence goals.

The first major issue that emerges from these experiences is that the two schools of architecture have proved to be sensitive to the most emblematic transformations that are investing their respective territories and have made of it the object of study. This is not surprising, given that the architecture is influenced not only by the techniques which are its own, but also by external forces that challenge its practice.

In the case of China, the massive movement "migration" of the rural population to the cities offers an unprecedented challenge that forces us to question the traditional techniques of construction of urban space and to establish new priorities: first and foremost, maximum energy efficiency and minimum emissions pollutants.

The European context also calls for a more efficient and sustainable cities with a focus on the territories of the margin, the peri-urban agricultural land threatened by misuse and degraded areas: the fate of these hybrids fields can play a decisive role on energy consumption and pollutant emissions management, but most are likely to affect resources critical to the autonomy of the city.

There are several occasions of overlap, but for an easier reading the contributions have been divided into four sessions. The first consists of the keynote' speeches. They introduce the main topics addressed in the following chapters, calling the suggestions from other cultural fields like in Cecchini one, or with D'Eramo contribution, stretching the concept of margin til reaching the ports as emblematic case where due to the container's revolution the city has broken its millennial relationship with its harbour.

The second session presents two study cases, placed in two very different context: an archipelago of hamlets in the North East of Sardinia (Italy) and a rural village in Ghana. In both situation the researcher explain the strategies adopted or planned to enhance the life quality of rural communities through actions that support their autonomy and through people empowerment, reinforcing the link between the communities and the agricultural field around the villages.

The third sessions addresses the future of peri-urban areas and relevant empty spaces spread into the consolidated city framework exploring the different planning strategies and tools to drive the development of those hybrids towards a more sustainable use.

The fourth and final session is devoted to strategies to reduce carbon emissions in a very challenging context such as China and particularly the regions of the North East. Here the conditions that affect pollutant emissions, already beyond the critical threshold, are subject to a surge in winter season, the climate exceptionally severe and prolonged that requires extraordinary energy resources to heat their homes and work spaces. Finally the Director of the EC2—Europe China Clean Energy Centre, presents the formalization of a partnership between the European Union and China aimed at promoting the use of alternative energy sources with operations in the territory of China that can model and to flying future programs as virtuous.

Alessandra Lai
Department of Architecture, Planning & Design-University of Sassari (Italy)

Contents

Part 1 Keynote Speech

The City and the Country: How to Reconstruct a Relationship

Arnaldo Cecchini[1,a], *Maria Rita Schirru*[2,b]
[1]*Department of Architecture, Planning & Design-University of Sassary (Italy)*
[2]*Department of Communication and Social Research-University of Rome "La Sapienza"*
[a] *cecchini@ uniss. it*; *b*[b]*mrita@ tiscali. it*

The greatest division of material and mental labour is the separation of town and country. The antagonism between town and country begins with the transition from barbarism to civilisation, from tribe to State, from locality to nation, and runs through the whole history of civilisation to the present day (the Anti-Corn Law League).

The existence of the town implies, at the same time, the necessity of administration, police, taxes, etc.; in short, of the municipality, and thus of politics in general. Here first became manifest the division of the population into two great classes, which is directly based on the division of labour and on the instruments of production. The town already is in actual fact the concentration of the population, of the instruments of production, of capital, of pleasures, of needs, while the country demonstrates just the opposite fact, isolation and separation. The antagonism between town and country can only exist within the framework of private property. It is the most crass expression of the subjection of the individual under the division of labour, under a definite activity forced upon him—a subjection which makes one man into a restricted town-animal, the other into a restricted country-animal, and daily creates a new the conflict between their interests. Labour is here again the chief thing, power over individuals, and as long as the latter exists, private property must exist. The abolition of the antagonism between town and country is one of the first conditions of communal life, a condition which again depends on a mass of material premises and which cannot be fulfilled by the mere will, as anyone can see at the first glance. (These conditions have still to be enumerated.) The separation of town and country can also be understood as the separation of capital and landed property, as the beginning of the existence and development of capital independent of landed property—the beginning of property having its basis only in labour and exchange.

This rightly famous passage from the German Ideology provides a key for the interpretation of the relationship between the city and the country in all its unbalance; undeniably, the city has never been "sustainable" in any sensible meaning of the word, we rather could say that the city is the place of organised human life with the paramount accrual of entropy.

Both the birth and the development of cities are under the sign of oppression of its exterior, of dissipation, of pollution, more in general of externalities.

The first great division of labour, the separation of town and country, condemned the rural population to thousands of years of mental torpidity, and the people of the towns each to subjection to his own individual trade. It destroyed the basis of the intellectual development of the former and the physical development of the latter. When the peasant appropriates his land, and the townsman his trade, the land appropriates the peasant and the trade the townsman to the very same extent. In the division of labour, man is also divided. All other physical and mental faculties are sacrificed to the development of one single activity. This stunting of man grows in the same measure as the division of labour, which attains its highest development in manufacture. Manufacture splits up each trade into its separate partial operations, allots each of these to an individual labourer as his life calling, and thus chains him for life to a particular detail function and a particular tool. "It converts the labourer into a crippled monstrosity, by forcing his detail dexterity at the expense of a world of productive capabilities and instincts."

Nonetheless, the city is one of the first fruits of the capacity of the mankind to control its future which has given rise to agriculture, and is at the basis, non only etymologically, of the civilisation (without forgetting though that culture is related to cultivation).

Closed urban form has been the rule until 700, limited by walls defending the city from the external world of the countryside, and beyond from the wilderness of the forest (wilderness is close to natural, while cultivated country is utterly artificial), places where the physical incolumity of the traveller could not have been protected and from where perils to citizens and the city could have come, in time of war and in time of peace.

The urbanlimes was defined, sharp, precise. At night the gateways were closed and outside the darkness took sway. The city gate had the same sacral value of the house doorsteps, the boundary between inside and outside, the barrier against the forces of evil.

Besides under the oppression, the city was also born under the sign of many marvellous things, but not unbound from the terrible ones.

We therefore need to take the city into account.

The problem of the limit has always existed, but perhaps it now almost always assumes not only local dimensions.

In the past, many cities, the predominant majority, have come to extinction for having destroyed their environment, above all their countryside and forests, that is, the conditions for their survival-essentially, an autophagia.

It is worth remembering that a characteristic of the modern city has been the breakaway from the "dogma" of spatial contiguity of the supplies and of the markets, while a characteristic of the contemporary city is the breakaway from the spatial contiguity of the production.

The urban expansion and the growth of urban consumption in the recent decades is determining also a globalisation of the environmental effects, both in terms of instantaneous impacts and of the stable pressure: not only the entire present world is invested, but also the entire future world is consumed.

Here in returns under many facets the original conflict between the city and the country.

There is a irreducible duality, a constant conflict, an opposition, between the two modalities of settlement of our species, that is, between the city and the country.

An opposition dividing the human settlements through history in distinct and complementary categories, a contraposition which in its purest form has had a well defined margin, a clear frontier, a sharp boundary in the city walls, almost always a necessary element of the city throughout history.

The boundary of city walls has always influenced the form of the city, increased its density to maximum levels, and contributed to the city's insalubrity.

But cities have often spilled over beyond the walls, faubourg were born close to the city gates.

Moreover, the city has often embraced agriculture within its walls, above all to resist sieges.

And foremost, daily, or rather nightly, a continuous flow of goods from the surrounding country flooded the general market, the abdomen of the city.

The general markets, such as Paris's Le Halles, the true protagonist of Zola's Le Ventre de Paris, were the point of encounter between the city and the country; at nights in the country the darkness rules and everybody sleeps, lays down in bed at sunset and wakes up with the first hours of sunrise, one can stay alone if one wishes; in the city the life is incessant and the nights are animated with lights and activities, pleasures and sins occupy theatres, cafés, pubs and streets: one could be able never to stay alone, as it happens in London to the protagonists of Poe's *The Man of the Crowd.*

This Is How Zola Talks About It:

Amidst the deep silence and solitude prevailing in the avenue several market gardeners' carts were climbing the slope which led towards Paris, and the fronts of the houses, asleep behind the dim lines of elms on either side of the road, echoed back the rhythmical jolting of the wheels. At the Neuilly bridge a cart full of cabbages and another full of peas had joined the eight waggons of carrots and turnips coming down from Nanterre; and the horses, left to themselves, had continued plodding along with lowered heads, at a regular though lazy pace, which the ascent of the slope now slackened. The sleeping waggoners, wrapped in woollen cloaks, striped black and grey, and grasping the reins slackly in their closed hands, were stretched at full length on their stomachs atop of the piles of vegetables. Every now and then, a gas lamp, following some patch of gloom, would light up the hobnails of a boot, the blue sleeve of a blouse, or the peak of a cap peering out of the huge florescence of vegetables—red bouquets of carrots, white bouquets of turnips, and the overflowing greenery of peas and cabbages.

And Poe:

As he proceeded, the company grew more scattered, and his old uneasiness and vacillation were resumed. For some time he followed closely a party of some ten or twelve roisterers; but from this number one by one dropped off, until three only remained together, in a narrow and gloomy lane, little frequented. The stranger paused, and, for a moment, seemed lost in thought; then, with every mark of agitation, pursued rapidly a route which brought us to the verge of the city, amid regions very different from those we had hitherto traversed. It was the most noisome quarter of London, where every thing wore the worst impress of the most deplorable poverty, and of the most desperate crime. By the dim light of an accidental lamp, tall, antique, worm-eaten, wooden tenements were seen tottering to their fall, in directions so many and capricious, that scarce the semblance of a passage was discernible between them. The paving-stones lay at random, displaced from their beds by the rankly-growing grass. Horrible filth festered in the dammed-up gutters. The whole atmosphere teemed with desolation. Yet, as we proceeded, the sounds of human life revived by sure degrees, and at length large bands of the most abandoned of a London populace were seen reeling to and fro. The spirits of the old man again flickered up, as a lamp which is near its death-hour. Once more he strode onward with elastic tread. Suddenly a corner was turned, a blaze of light burst upon our sight, and we stood before one of the huge suburban temples of Intemperance—one of the palaces of the fiend, Gin.

The duality city/country brings along that of night/day, solitude/multitude, light/darkness. But, we want to emphasise, not that of natural/artificial.

In different phases of history, this duality has been feeding the constant resistance of the country to the domination of the city, a resistance which once was also of the countrymen, and now after the abandonment of much of the countryside (the city and the forest win over the agriculture, so say the data), is substituted by the force and the viciousness of the "nature", by the sheer force of its entropic processes, and by its irreducible power.

The country and the city, the two artificial forms of organisation of the territory, have put in discussion the harmony, spontaneous and natural, between the human beings and the nature in a way present before the discovery of the agriculture: as we have said, they both are the expression of the capacity of domination of our species, of its capacity to imagine and construct the future.

The country and the city, two entities forced to coexist, even if in some cases the country can live without the city, while the city, until recently, could almost never have lived without its country.

Two Forms of Life, Two Ways of Thinking, Two Cultures.

From one side, the contempt of the citizens for the peasants and for the rustic. A concrete, material, permanent contempt, a contempt shrouded in one of the most famous texts of the history:

The bourgeoisie has subjected the country to the rule of the towns. It has created enormous cities, has greatly increased the urban population as compared with the rural, and has thus rescued a considerable part of the population from the idiocy of rural life. Just as it has made the country dependent on the towns, so it has made barbarian and semi-barbarian countries dependent on the civilised ones, nations of peasants on nations of bourgeois, the East on the West.

The envy of inhabitants of the city for the country, but often for a false, imaginary country. A literary and inconstant envy.

The country, as opposed to the city described as gasping and anguishing nightmare, is no less than a reference to other worlds, colonised and distant, places of hope for a humanity lost, deserving punishment, according to different points of view of the future of the planet. We can in this sense say that the division between the city and its rural surrounding gets transferred into the space, replicating an ancient and unresolved dichotomy, already presented by Aesop in the VI century BCE in the contrast in the ways of life between the city mouse and the countryside mouse; already then, the life in cities was associated to the angst which seems to assume even an existential dimension, emphasised by the traditional diffidence of the countrymen for the city and its attractive force.

Let Us Recall Horace's Version of the Fable:

Then says the rustic: "It may do for you,
This life, but I don't like it; so adieu:
Give me my hole, secure from all alarms,
I'll prove that tares and vetches still have charms."

The urbaphobia among urban utopias, almost always frightened by the constant expansion of the city, by the progressive expulsion of any agricultural activity, of any undeveloped land, of any green area, imagines-unable to do without the city, but not loving it (the *ville mal aimèe* is a constant in a part of the nineteenth-century literature—small cities, "besieged" by the countryside, self-sufficient, in harmony with the context.

The factory town, however, transforms all water into stinking ditch water. However much therefore concentration in the towns is a basic condign of capitalist production, each individual industrial capitalist is constantly striving to get away from the large towns necessarily created by it, and to move towards exploitation in the countryside.

Once more, only the abolition of the capitalist character of modem industry can bring us out of this new vicious circle, can resolve this contradiction in modern industry, which is constantly reproducing itself. Only a society which makes possible the harmonious co-operation of its productive forces on the basis of one single vast plan can allow industry to settle in whatever form of distribution over the whole country is best adapted to its own development and the maintenance of development of the other elements of production. Accordingly, abolition of the antithesis between town and country is not merely possible. It has become a direct necessity of industrial production itself, just as it has become a necessity of agricultural production and, moreover, of public health. The present poisoning of the air, water and land can only be put an end to by the fusion of town and country; and only this fusion will change the situation of the masses now languishing in the towns, and enable their excrement to be used for the production of plants instead of for die production of disease.

The description of the city, dirty, dangerous, immoral, insalubrious, dissipating (not distant from true), in contrast to the country, tidy, peaceful, immaculate, healthy, sober (totally false), is at the basis of many products of science fiction; there are plenty of cinema references inspired by sulphurous, oppressive visions, almost always after a nuclear or an environmental catastrophe, such to profoundly modify the relationship between citizens and institutions turned authoritarian and all-present in order to control a society devoid of any cohesion: Orwell's 1984, Scott's *Blade Runner*, but also the "minor" films (like various *terminators* and other ersatz productions), as if to demonstrate the existence of cities no longer urban places but indistinct containers of humanity indifferent to its own destiny.

This contrast is at the basis of the so called Asimov's urban utopias, described in three of his science fiction thrillers *The Caves of Steel*, *The Naked Sun* and *The Robots of Dawn*. In the trilogy the becomes even more extreme, with the whole Earth as one great city, and external planets colonised by men more or less radically only as country.

"It can just be made out in the gap between the two Brunswick Sectors. Low domes spread out. It's the difference between us and the Spacers. We reach high and crowd close. With them, each family has a dome for itself. One family: one house. And land between each dome. Have you ever spoken to any of the Spacers, Lije?" "About a month ago, I spoke to one right here on your intercom". Baley said, patiently. "Yes, I remember. But then, I'm just getting philosophical. We and they. Different ways of life".

It could lay itself out scientifically. At the center was the enormous complex of administrative offices. In careful orientation to one another and to the whole were the large residential sections connected and interlaced by the expressway and the local ways. Toward the outskirts were the factories, the hydroponic plants, the yeast-culture vats, the power plants. Through all the melee were the water pipes and sewage ducts, schools, prisons and shops, power lines and communication beams.

The radical change had been the gradual formation of the cities over a thousand years of Earth's history. Efficiency implied bigness. Even in medieval times that had been realized, perhaps unconsciously. Home industry gave way to factories and factories to continental industries. Think of the inefficiency of a hundred thousand houses for a hundred thousand families as compared with a hundred-thousand-unit section; a book-film collection in each house as compared with a section film concentrate; independent video for each family as compared with video-piping systems. For that matter, take the simple folly of endless duplication of kitchens and bathrooms as compared with the thoroughly efficient diners and shower rooms made possible by City culture.

Each city became a semiautonomous unit, economically all but self-sufficient. It could roof itself in, gird itself about, burrow itself under. "It became a steel cave, a tremendous, self-contained cave of steel and concrete."

In primitive times, individual population centers were virtually self-supporting, living on the produce of neighboring farms. Nothing but immediate disaster, a flood or a pestilence or crop failure, could harm them. As the centers grew and technology improved, localized disasters could be overcome by drawing on help from distant centers, but at the cost of making ever larger areas interdependent. In Medieval times, the open cities, even the largest, could subsist on food stores and on emergency supplies of all sorts for a week at least. "When New York first became a city, it could have lived on itself for a day. Now it cannot do so for an hour. A disaster that would have been uncomfortable ten thousand years ago, merely serious a thousand years ago, and acute a hundred years ago would now be surely fatal."

A space between two trees revealed an expanse of lawn. For the first time, there was a sense of distance and on the horizon one could see a dwelling place: low-roofed, broad, and so green in color that it almost melted into the countryside.

"This is a residential area." said Fastolfe. "It might not seem so to you, since you are accustomed to Earth's tremendous hives, but we are in the Auroran city of Eos, which is actually the administrative center of the planet. There are twenty thousand human beings living here, which makes it the largest city, not only on Aurora but on all the spacer worlds. There are as many people in Eos as on all of Solaria."

"How many robots, Dr. Fastolfe?"

"In this area? Perhaps a hundred thousand. On the planet as a whole, there are fifty robots to each human being on the average, not ten thousand per human as on Solaria."

"What happens as the population rises? Do you add more robots?"

Fastolfe shook his head. "The population does not rise."

"Aurora's population is two hundred million and that has remained stable for three centuries. It is the number desired. Surely you have read that in the books you viewed."

"Yes, I have," admitted Baley, "but I found it difficult to believe."

"Let me assure you it's true. It gives each of us ample land, ample space, ample privacy, and an ample share of the world's resources. There are neither too many people as on Earth, nor too few as on Solaria."

Aurora is mentioned in the following extract (and briefly Solaria); note the similarity with the "garden city" project:

"We're passing through its center. The limits are seven kilometers away and our destination is nearly forty kilometers beyond that."

"The center of the city? I see no structures."

"They are not meant to be seen from the road, but there's one you can make out between the trees. That is the establishment of Fuad Labord, a well-known writer."

"Long distances are covered by air-cars or magnetic subcars. Tridimensional connections—"

"They call it viewing on Solaria." said Baley.

"And here, too, in informal conversation, but TVC more formally. Finally, Aurorans are fond of walking and it is not unusual to walk several kilometers for social visiting or even for business meetings where time is not of the essence."

"And we have to get somewhere that's too far to walk, too close for air-cars, and tridimensional viewing is not wanted...."

A dualism which in Asimov almost results in a sort of a speciation; one seems to get a glimpse of the same possible speculation made by Aberto Salza about the growing divide in wealth and power.

The ruraphobia instead does not need to find forms of expression. Rather than a proper phobia it is a disinterest, a distraction, an indifference, the fact that the exploitation of the country and the dominion over it is considered banal and obvious.

The bottom line of the urbanisation in the last decades is not only the increasing expansion of the city (for the first time in the history overtaking the country in terms of the living population), the most significant phenomenon, as we already underlined, is the progressive disappearance of the country: the land is divided between built and abandoned or undomesticated, while the agricultural land gets reduced; in effect, one of the two poles of the dialectic of anthropic spaces is disappearing, brutally pitting the city against the manless nature.

The power of nature and its indifference to the survival of our and all the other species is wonderfully staged in Alan Weisman's book *The World Without Us*: a sudden and complete disappearance of our species unleashes the forces of nature and time to re-conquer and re-mould the anthropic spaces, we could call it the come to power of the "third landscape".

The duality city/country is the duality between two forms of domesticated nature, both artificial, both planned (paradoxically, more of the country than of the city and of its invasion outside itself), an always conflicting dialectic, always under the sign of the oppression of the city over the country, but a dialectic which has sometimes produced equilibrium and development (not always and not everywhere, it should be remembered) while they lasted.

But if one of the poles of the duality disappears, if the couple splits, then the entire order of the territory becomes fragile, so much so that a barely consistent rainfall provokes floods, landslides, disarray.

It is a double from which to start anew for those who deal with the government of territory.

And it is not out of naivety of generalisations about the agriculture of proximity or zero kilometre, or out of the diffidence for a heedful use of technologies in the agricultural production processes.

The food self-sufficiency of territories (defined how? how big?) is in itself a nonsense, and in contract with the history.

Capitalist industry has already made itself relatively independent of the local limitations arising from the location of sources of raw materials. The textile industry, in the main, works up imported raw materials. Spanish iron ore is worked up in England and Germany and Spanish and South American copper ores are used in England. Every coalfield now supplies fuel to an industrial area beyond its own borders, an area which is widening every year. Along the whole of the European coast steam engines are driven by English and to some extent also by German and Belgian coal Society liberated from the barriers of capitalist production can go much further still.

But that the city-putting aside very rare exceptions—needs its country, a well defined and organised area of cultivated and looked-after territory, and that from an economic point of view this necessity gives an important space to the agriculture of proximity, appears—at least from the point of view of the government of territory-essential and relevant also for the technological experimentation, for the conservation of diversity, for the culture.

We can claim, without nostalgic concessions, that the intrusion of the country within the city (what is known as urban farming) from an environmental and urban planning point of view is an effective response to the need for green areas, for the reduction of pollution and of greenhouse gases.

The glorious survival of the city, which after allis the ecological niche of the human species, will not be possible if we give way to the annihilation of the country: the *reductio ad unum* of the forms of anthropic space is the premise (and the consequence) of the crisis of the city, related to another *reductio ad unum* which is its *causa prima*: the acceptance of the *pensée unique* as the sole criterion for the government of cities.

Rather than passing from the double to the one, we should consider the idea to let ourselves be seduced by a third protagonist, that which has appeared at some point in this reflections, that which Gilles Clément in an inspiring intuition called "the third landscape":

If one ceases to regard the landscape as the object of an industry, one immediately discovers—is it an oblivion of the cartography or a negligence of politics? a heap of undecided spaces, devoid of functions difficult to name. This set of spaces belongs neither to territory of the shadow nor to that of the light. It is situated on the margins.

At the borders of the woods, along the roads and rivers, within the forgotten nooks of cultivations, where the cars cannot pass. Spaces of modest dimensions, scattered like the lost angles of a field; uniform and wide as peatlands, moors and certain recently deserted fallow lands.

No resemblance of form within these fragments. Only one point in common: they all are a territory of refuge of the diversity. That which is dispelled from elsewhere.

This justifies to collect them together under a single term. I propose third landscape, the third term of an analysis that places the main elements below the shadow on one hand, below the light on the other.

Third landscape refers to the Third Estate (not to the third world). A space that expresses neither power nor submission to power.

To make reference to the Sieyès's pamphlet in 1789: "What is the Third Estate? Everything. What has it been until now in the political order? Nothing. What does it ask? To become something."

We could think that instead of the frontier, of the *limes*, we should think of a border space, a frontier: a margin, an undecided space; instead of the disappearance of the duality, the reconstruction of a plurality, or better, of nested dualities.

Another duality resides in the relationship between the anthropized environment (city and country) and the abandoned and derelict environment; the abandoned zones are vital, necessary, as demonstrated by the border stripe between two Koreas, a 907 square kilometres demilitarised zone (248 by 3 and a half km), which since more than seventy years is completely devoid of humans, at total disposal of the third landscape, an incredible concentration of biodiversity.

The second duality resides in the relationship between the two main modalities of anthropized spaces, urban and rural, which should well not be separated by a sharp boundary, but should have permeable frontiers open to reciprocal intrusions, reconstituting the interacting duality of the *Tajiutu*, a dynamic, complementary and mutually contaminating double.

We have talked about the limit of the city, that physical, psychological and symbolic of the city walls.

With the evolution of military techniques, with the birth of the nation states, with the specialisation of production, with the development of commerce among cities and among states, the urban perimeter, the walls, the border slowly perish and lose its functional and symbolic value.

In the contemporary world the border loses form and becomes substance.

"Terminus, or term -en, -inis, that is termo, -onis derives from the root ter which means 'to cross', 'to reach a destination laying beyond'. Hence terminus essentially indicates a border. Originally this border was conceived as materially traced, and therefore the most significant kernel of the words belonging to this etymological family indicated a border pole, a border rock, a sign of the border in general. In Greek, the word terminus, much in philosophy as in wider use, corresponds to the word xopoc, a 'circumscribed place' or a 'place for dance', and furthermore the word xorismos, 'separation', 'difference', 'isolation'. This etymological analysis shows, then, that the term is originally a custodian of the border of the colture/culture; it brings to life articulation and structure, it firmly fixes its fundamental articulations and, by not permitting a general confusion, while limiting the life in reality it liberates it towards a further creation. And the same term, in so far as a border of a given region of culture, belongs to that culture. I will use a mathematical expression: the term is the limit meaning of the culture. The term is a border my mean of which the thought self-delimits itself, becoming aware of itself."

In mathematics, the limit is the bearer of a principle of infinity. To draw a limit to infinity becomes in our discourse the metaphor of multiplying the existential modalities foreign to the living in the city.

It means to recover the natural horizon of perception, to see the sunset, to widen the space seen and lived as far as the eye can see, to free ourselves from the architectonic barriers of the urban landscape and to permit the soul to retake possession of the world.

It is to describe a path across the border, a mysterious and poorly frequented place. The place we often encounter in our movements, the place where it is easy to bump into the unexpected and to move, often gropingly, in distress. We would like, that is, to begin to observe the strange space found "within" the things, that which separates when put in contact, or, perhaps, puts in contact when separated, persons, things, cultures, identities, spaces different among each other. Thus, a space of border, but also (it is at least an hypothesis) a border as a space. The space can have an external margin, that where man lives, but also an internal margin, interior, intimate, linked to our states of mind, and to hopes and utopias accompanying them. "Margins which we rarely are able to clearly observe, even if we affirm their existence with certainty."

The suburban, the outpost of the *limes*, is nonetheless a "middle-earth" where the areas of margin, of interstice, of forgotten void, can lower the quality of life and relegate it to melancholic peripheries, which are nothing less than the new city walls of the spirit.

But it could be otherwise.

The areas of border, the frontiers, are possible places of mediation among different and in some cases antagonist forces: through mediations we find new values and new forms to better structure the future. The natural and agrarian conformation of the landscape interacting with the urban limes should activate a "reverberating" effect, blending and integrating existential models which in their diversity complete and harmonise each other.

Thought of in this way, the geographical *limes* loses its characteristic of the "limit", of a separating border, and becomes a way to open to other spaces and other places.

"But the border line, as we have had ways to emphasise, always presents a double aspect and meaning: it is on one hand a line of demarcation and separation, on the other the place of contact and of mutual exchange among different spheres which, while staying separated in their specific determination, can enter in communication precisely through this trait d'union. From this second point of view it presents itself as the place of the specific contact between internal and external."

To sum up: the frontiers, the areas laying around the border, are structural elements with the value of margin between the urban and the agricultural landscape.

They are portions of the territory which blend elements belonging to both places. They are zones of conflict, but highly dynamic as mediators of possibilities. The urban fabric is rarefied and tends towards de-structuring and the loss of form. The fragmentation, the discontinuity, can relegate it to insignificance, and its rapid transformation to the impossibility to manage the proper dynamic of the urban *limes*.

These aspects are strongly conditioned by the functional components and by the modalities of fruition of the territory.

It would be necessary to examine in more depth the socio-economic aspects which always mould the becoming of the city, integrating them with the quality of the interventions of restoration and rehabilitation of the territory, intended as the landscape-environmental compartment around the urbanised, formed and stable settlement.

The urban analysis of the *limes* and of its dynamic should be the occasion to recover quality and beauty.

It is a challenge of extreme complexity: the urban explosion often manifesting itself as *sprawl*, but isn't only *sprawl*, is the modern version of the oppression of the country by the city, an oppression with destructive pressure seen many times in the past, in many ways, but almost always with catastrophic effects.

Nonetheless, no rare are cases in which the city and the country in the processes of peri-urbanisation were capable of finding a positive bond whose most relevant aspects are represented by the extension of the areas of concentration of the workforce, by the nearing of farmers to the city, by the integration among activities and between those and the environment, etc..

City lives, with the entirety of different functions which characterise it with respect to any other type of community and settlement, it lives with its conflicts, its colloquialisms, its inefficiencies, its dissipations, its relative unsustainability.

It lives though for it is capable of being relatively sustainable, that is to say, to assure in the same time its present, its past and its future, itself, its countryside and its environment, to manage its limits; not all cities of the past were capable of that, as Venice was for centuries (see the wonderful book *Venice and the Waters* by Piero Bevilacqua).

The disappearance of Venice as city—by now rather near and even probable—will be due to both of these pressures: on one hand the incapacity to control and limit the pressure of the city on its environment (among others due to neglect and carelessness), on the other hand the loss of social functions and of the diversity, the full and uncontrolled supremacy of the touristic monoculture over all the activities, greedy for resources and unsustainable as all the monocultures, insatiable in devouring spaces and people.

How to manage devouring famishment of cities, whether compact and overpopulated, distributed, dispersed, polycentric, or the possible "low-density cities" of the great parts of Sardinia, is always a problem; so much more in an epoch in which not only the cities host the majority of the mankind, but in an epoch in which the environment of the city is not only or prevalently local as it almost always used to be in the past, but truly global and therefore the pressure on its limits is in good part governed and determined from the outside.

The impacts of the famish cities used to be only local: today for the first time in history the urban explosion everywhere invests the surrounding territory, heading to a potential global catastrophe. To cope with the phenomenon of urban explosion is not simple also because, as we said, coupled with the pressure of building development on the country; the solution, perhaps, isn't a return to the compact city, but to consider some phenomena of urban diffusion as the seeds for a possible rural urbanity, to be accompanied with the seeds of urban rurality.

The interstitial spaces, the abandoned areas of the cities need not necessarily be used, even if often to re-use them is an essential alternative to avoid the land occupation for building development: they could also usefully become pieces of country or, also within the city, left abandoned for the processes of re-naturalisation to take over.

The two is instable, the spontaneous entropic tendency is towards one; to maintain the duality is the result of a projection towards the future, in fact, of a project. The project is what the city and the country need, what the landscape need.

We hence encounter, once again, a border territory, to be explored and in which to meet free of constraints and of conditionings imposed by a too marked adhesion to consolidated habits and to the most recurrent schemes of thought. I hope that the suggested itinerary could induce to dwell a bit "on the threshold" of the problems, in a hesitation which refutes to take everything for granted and which can, precisely for that reason, transform into the demand for the necessity to safeguard a plurality of dimensions. In the moments of crisis and difficulty, of theoretical dismay, one ought better to behave as Alexander Pope's angels, hesitant to put their feet in territories where, vice versa, the unwise burst into, gay and reckless.

REFERENCES

[1] MARX K, ENGELS F. The German ieology [M/OL]. 1845-1846. http://www.marxists.org/archive/marx/works/1845/german-ideology/index.htm.

[2] ENGELS F. Anti-dühring herr eugen dühring's revolution in science[M/OL]. 1877. http://www.marxists.org/archive/marx/works/1877/anti-duhring/.

[3] DIAMOND J. Collapse: how civilities choose to fail or succeed[M]. New York: Viking Press,

2005.

[4] ZOLA E. The fat and the thin; or, the belly of Paris[M/OL]. 1873. http://books.google.it/books? id=3VEMAoJrCmkC&printsec=frontcover&hl=it&source=gbs_ge_summary_r&cad=0#v=onepage&q&f=false.

[5] POE E A. The man of the crowd[M/OL]. 1850. http://xroads.virginia.edu/~hyper/POE/manofcro.html.

[6] MARX K, ENGELS F. Communist manifesto[M/OL]. 1848. https://www.marxists.org/archive/marx/works/1848/communist-manifesto/ch01.htm.

[7] ASIMOV I. Caves of steel[M]. New York: Doubleday, 1954.

[8] ASIMOV I. The robots of dawn[M]. New York: Doubleday, 1983.

[9] SALZA A. Niente: come si vive quando manca tutto-antropologia della povertà estrema[M]. Milan: Sperling & Kupfer, 2009.

[10] ZANINI P. Significati del confine: i limiti naturali, storici, mentali, Bruno Mondadori[M]. Milan: Milano Press, 1997.

[11] SCHIRRU M R. Il periurbano: crescere intorno alla città[M]. Roma: Gangemi Editore, 2012.

The Knowledge Cities: How the Universities and Science Parks are Shaping the Chinese Urban Environment

Plinio Innocenzi[1,a], *Maurizio Minchilli*[1,b]

[1]*Department of Architecture, Planning & Design-University of Sassary (Italy)*

[a] *innocenzi@uniss.it*; [b]*minchilli@uniss.it*

The urbanization in China is an unprecedented phenomenon in terms of dimension and rate and represents a tremendous challenge for the complexity and the scale the urban planners have to deal with. Urbanization has been identified by the Chinese government as the main tool to reduce the income gap which is particularly high between rural and urban population.

The increase in urban population in China is a quite linear phenomenon, at least in the last 12 years; this is very likely the effect of centrally planned policies which have tried to monitor and control, within some limits, the movement from rural areas to the cities. The urbanization rate has remained quite constant with an increased of urban population increased from 39.09% in 2002 to 53.49% in 2012; in 2010 the rural population has paired the one in the cities, which from 2011 became larger (Figure 1).

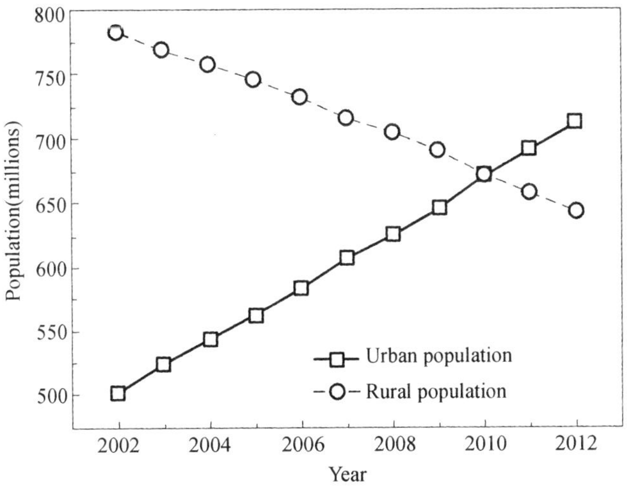

Source: Statista 2013

Figure 1 Variation between urban and rural population from 2002 to 2012

The linear increase allows forecasting that on 2025 the urban population will reach one billion people with 221 cities with at least one million people and other 23 whose population will exceed 5 millions. All these cities will generate 90% of the Gross Domestic Product (GDP) of China, becoming the economic and financial core of the country (McKinsey Global Institute). Some cities will reach the dimension of megacities with a population of more than 10 million people: Beijing, Shanghai, Tianjin, Shenzhen, Wuhan, Chongqing, Chengdu and Guangzhou. On 2025 other 350 million Chinese will be added to the urban population, which is around the actual size of United States.

These megacities will generate the main GDP growth from now up to 2025 in comparison with the other cities of China and if we look to the global forecast, 5 Chinese cities will be within the top 25 in terms of population and 14 as GDP growth (Global Metro Monitor Volatility).

In this frame of continuous growth of the Chinese megacities a fundamental role is played by the centres of knowledge such as universities, science parks, science cities and so on. The capability of attracting the most creative and talented people is very much linked with the presence of these clusters, whose dimensions has to be properly managed as a function of an increasing demand in terms of access and quality(Table 1).

The availability of qualified human resources is a fundamental need for creation of wealth within the cities, such as the presence of the technological clusters as science parks and high tech development zones (Siddivò M.).

The Chinese planners have widely used these instruments for guiding and boosting the cities economic growth through the creation of advanced research platforms for industries and academia which are closely connected with universities and technical vocational schools. These centres are used to attract domestic and foreign investments but also to create the critical mass in terms of human resources and infrastructures to support the urbanization via creation of new jobs and new industries.

There is no a single model which has been applied as a frame tool for creation of such knowledge clusters, and in general local governments and planners have had enough freedom to adapt the system to the local specificities. An example of such flexibility is the programme Torch (The China Quarterly) which is planned at central level but implemented at the provincial or county level with very much flexibility.

Table 1 The world megacities in 2025, rank in terms of GDP, GDP growth and number of high income families (at least 20 000 $)

Rank	GDP	GDP growth	Population	High income families
1	New York	Shanghai	Tokyo	Tokyo
2	Tokyo	Pechino	Mumbai	New York
3	Shanghai	New York	Shanghai	Londra
4	Londra	Tianjin	Pechino	Shanghai
5	Pechino	Chongqing	Delhi	Pechino
6	Los Angeles	Shenzhen	Calcutta	Parigi
7	Parigi	Guangzhou	Dakka	Rhein-Ruhr
8	Chicago	Nanjing	San Paolo	Osaka
9	Rhein-Ruhr	Hangzhou	Città del Messico	Mosca
10	Shenzhen	Chengdu	New York	Città del Messico
11	Tianjin	Wuhan	Chongqing	Los Angeles
12	Dallas	Londra	Karachi	San Paolo
13	Washington D. C	Los Angeles	Kinshasa	Seul
14	Houston	Foshan	Londra	Chicago
15	San Paolo	Taipei	Lagos	Milano
16	Mosca	Delhi	Cairo	Mumbai
17	Chongqing	Mosca	Manila	Cairo
18	Randstad	Singapore	Shenzhen	Hong Kong
19	Guangzhou	San Paolo	Los Angeles	Taipei
20	Città del Messico	Tokyo	Buenos Aires	Randstadt
21	Osaka	Shenyang	Rio de Janeiro	Shenzhen
22	Filadelfia	Xian	Tianjin	Istanbul
23	Boston	Dongguan	Parigi	Nuova Delhi
24	San Francisco	Mumbai	Giacarta	Buenos Aires
25	Hong Kong	Hong Kong	Istanbul	Madrid

1 UNIVERSITIES AND CITIES IN CHINA

A fundamental role in Chinese urbanization is played by the Universities, indeed for a double reason, the first one for their traditional role to form the human resources and a second one, which is very much peculiar of China, as a direct player of urban growth.

The Chinese universities form a highly hierarchical and rigid system, with a national rank that is continuously updated and implemented. At the top of this pyramid there are the 39 Universities of the 985 programme which together with the 211 group (the 100 universities for the 21th century) form the elite of China education system. These top universities attract most of the investments in education and research with 90% of the Key State National Laboratories and 70% of global research funds. The 985 project has been dedicated to the creation of research universities based on the American model (the 39 top Universities in the rank of 211).

There is main drawback in this education system that is the not homogeneous distribution within the country that on the other hand has the dimension of a continent. There is a strong polarization in the areas of Beijing and Shanghai with an impressive concentration in the Capital city, 8 of 985 Universities, the double of Shanghai and 29 of 211 Universities with respect to the 12 of Shanghai. To these universities have to be added the main Academies, the most important Chinese research Institution is the Chinese Academy of Sciences (CAS), then the Chinese Academy of Engineering, the Chinese Academy of Social Sciences, the Chinese Academy of Agricultural Sciences, the Chinese Academy of Forestry just to cite some, and all of them have the headquarter in Beijing and the larger part of their research capability.

The education pyramidal system based on the American model of research and teaching universities and the strong polarization in term of location in the country has very much contributed to the formation of knowledge cities, such as Beijing and Shanghai. On the other hand large provinces have almost no access to education of quality with the result that the Chinese students are forced to emigrate abroad looking for the possibilities of getting a good education which do not have at home.

The hukou system, which links the population to their birth place with few possibilities to be integrated in other cities contribute also to this polarization, because the larger part of the places in the best universities are reserved to students with local hukou. The result is that the two main cities of China have become such a strong concentrators and attractors of talents and high-qualified human resources with respect to other provinces of China.

The Chinese Universities are all based on the Campus model and they form, with few exceptions an isolated environment within the cities, some students very seldom get out of their compounds. These Campus are almost independent and self sustainable small cities with residence and services for all the population of professors and students replicating the Mao danwei production units which were completely independent with few interactions with the outside world. These Campus-danwei have been originally localized at the border of the main cities, but with the fast growth of the last years with urbanization have been incorporated within the core of the cities and now, such as the case of Beijing are in the main urban area, at the centre of the Haidian district (Table 2). The centre of the city is divided in two districts, Xicheng and Dongcheng, around them are localized other four urban districts with 6 outer districts; at North there are other two districts and two rural counties. The Zhonguancun area is within the Haidian district and is expanding to the outer district of Changping.

Table 2 The 985 Universities in Beijing, Shanghai and Tianjin

Municipality	985 University
Beijing	Tsinghua University
	Peking University
	Beijing Normal University
	Beijing Institute of Technology
	Renmin University of China
	China Agriculture University
	Minzu University
	Beijing University of Aeronautics and Astronautics
Shanghai	Shanghai Jiao Tong University
	Fudan University
	Tongji University
	East China Normal University
Tianjin	Tianjin University
	Nankai University

The incorporation of the Campus within the cities has created a series of problems, because the expansion of the Universities, which traditionally have a relatively small number of students, between 20,000 and 30,000, is physically hampered by the growth of the city. The response is the creation of satellite Campuses in the outer areas of the city, which on the other hand strongly contributes to the expansion of the city. It is a phenomenon which is from one side triggered by the need of the University but also by the real estate speculation which has always a strong role in governing the urbanization in China.

The result is the increase of urban sprawl and the consumption with an apparently endless expansion of the city border.

The new Campus of the University of Science and Technology of Beijing is located around 40 km from the main one, the same for the new Campus of Shanghai Jiao Tong University and Tianjin University. The distances are too big to be covered by daily commuting and the professors and their families are forced to relocate in new towns which have been built close to these new Campuses. The economic interest behind such real estates speculation is so strong to form an abnormal twine between universities administrators and developers. The presence of the high level university produce an immediate increase of the prices of all the houses in the area which is a formidable boost to promote the city expansion.

2 BEIJING AND THE CHINESE SILICON VALLEY

The case of Beijing is particularly interesting because the Capital city has become in a few years one of the main centres of knowledge in the world and a strong driver for innovation, especially in the sector of ICT. Until the opening of Deng Xiaoping the Capital city was shaped by the Sovietic planning followed by Chairman Mao, the urban panorama was filled by heavy industries, still factories, chemical and mechanical plants.

The Russian counsellors suggested that the industries should be located mainly in the Haidian district (Figure 2). The new era of opening completely reshaped this urban structure and the industries were moved to the outer areas or neighbouring provinces, the universities also changed their mission and transformed from specialised to general one. The expansion of the city brings a fast deindustrialization process with the fast growth of services and research activities. The district of Haidian was interested by an impressive transformation, from heavy industries to a cluster of universities, research centres and ICT small companies. From 1999 the area has been officially named Zhongguancun Science & Technology Zone (ZSTZ) (Zhou Y.) and is composed of seven science and technology parks, 27,000 enterprises with 21,000 identified as high-tech and 3,600 start up. In the same area are located 15 between the best universities of China and some leading companies such as Lenovo which is a spin off of the Chinese Academy of Sciences and Baidu a leading company of Internet.

Figure 2 The districts of Beijing Municipality

This concentration of knowledge in the high tech ZSTZ is unique in China, all the chain of creation of value and production is based in one well defined area of the city and is centrally planned and controlled at least in the main expansion lines. This location has also the advantage to reduce the consumption of energy and optimizing the effect of clustering high tech companies in an urban area. On the other area the growth of the city has been strongly shaped by the presence of this central knowledge and high tech production cluster which has become a compelling attractor of talents forming a self-propelled virtuous circle. At the same time the Zhongguancun has favoured the growth of a dense city limiting the urban sprawl; this scenario could quickly changes in a next future because of the vertiginous rise of housing and living costs that could force the rise of new development areas in outer areas of the city.

3 THE SHANGHAI MODEL, DELOCALIZATION OF KNOWLEDGE

Shanghai is the only city of China which has a university system that can be compared to that one of Beijing. The development model of Shanghai is however deeply different and in comparison to the Capital city the growth of the city has been favoured by a dispersion of the knowledge centres more than clustering. The parks and universities are quite far from each other and are mostly located in the suburbs or outer new expansion areas more than downtown where has grown a financial and commercial area (Table 3).

The new campus of Shanghai Jiao Tong University, such as the new biotechnology science park are located at more than one hour by car from the Bund area in the city center.

Table 3 Distance between the main universities of Shanghai and the science parks

Location			Distance
From		To	km
University Shanghai Jiao Tong	Xuhui Campus	Zhangjiang High Tech Park	19.10
University Shanghai Jiao Tong	Xuhui Campus	Fengjiang Industrial Zone	35.70
University Shanghai Jiao Tong	Campus Minhang (Main)	Zhangjiang High Tech Park	41.00
University Fudan		Fengjiang Industrial Zone	60.60
University Fudan		Zhangjiang High Tech Park	16.50
Songjiang Industrial Zone		Fengjiang Industrial Zone	48.40
Songjiang Industrial Zone		Zhangjiang High Tech Park	49.10
Zhangjiang High Tech Park		Caohejing High Tech Park	37.80
Zhangjiang High Tech Park		Fengjiang Industrial Zone	50.10

Source: Breznitz D., Murphree M.

The main consequence of this choice is that the main core business of the city is given by heavy state industries, heritage of the Sovietic model; they form an industrial belt around Shanghai which has somehow limited the rise of the small innovative companies in the ICT sector which are found in Beijing. The dispersion of research centres and universities does not favour the small enterprises, even some changes can be envisaged such as the born of a large biotechnology pole where are going to be concentrated most of the research activity in the field.

The other drawback of this model is the limited central planning of the activities, most of the parks are in fact managed at local level which means that competition, overlapping of functions can not be avoided. In any case this situation triggers the expansion of the city to the outer area, in a continuous soil consumption and urban sprawl which create the conditions for increasing pollution and traffic problems.

4 THE THIRD LINE, CHENGDU THE HIGH TECH CORE OF CHINA

The city of Chengdu, the capital of Sichuan in the South-West of China, is another interesting case to be examined and is also one of the most successful because in a quite short time the rise of a huge high tech park has completely reshaped the destiny of the city.

More than 200 of the top 500 company in the Fortune list have opened an activity in Chengdu; this success is in great part the result of a careful planning which starts from the implementation of the Torch programme at local level. The management model has followed a path similar to that one of Beijing more than Shanghai. The Chengdu Hi-Tech Development Zone has been divided in two areas, North and South, and the city has grown following concentrical rings (ring roads).

In the South has been created the new of Tianfu which is linked with Chengdu with a mobility main road and a subway; in the West area is localized the Chengdu West Hi-Tech Park and in the South the Chengdu South Hi-Tech Park. In Chengdu have been clusterized companies which are specialized in specific strategic sectors, such as ICT, bio-pharmaceutical, opto-electronic, microelectronics, flat screens.

To support the industrial growth has been also created an education system which is able to give the high qualified human resources for the high-tech industries; 42 colleges and universities with 590,000 students, 100,000 undergraduate students are formed every year with 50,000 in the science and technology and 10,000 with a graduate degree.

The research infrastructure is formed by 70 research institutes, 31 company technology-centres, 37 sections of the key state national laboratories. The system does not form highly specialized top research but the qualified manpower at an intermediate level, which is necessary for the companies localized in the science and technology park.

The Park is well integrated in the urban area, which has largely been planned on the basis of the growth of the science park.

The High Tech Development Zone in Chengdu has surely reduced the urban sprawl and has been an attractor of qualified human resources and talents. The specialization, the central planning and the qualification of the man power are at the root of the success. Of course the enormous growth of the city has not been done with zero costs, the consumption of soil is enormous and the city is expanding at a rate which is even going beyond the most optimistic expectations.

5 CONCLUSIONS

The previous examples of Beijing, Shanghai and Chengdu clearly show that the universities and science parks represent one of the main triggering force in the urbanization process of China. In the Table 4 are listed the main characteristics and differences of the three different models and the effects on the city with respect to the High Tech Development Zone. These zones have catalysed the urban growth and transformation being an integrated tool for planners and developers. They have played a multiple role, as attractors of talent, as a pool of innovation and as a driver for property speculation. There are very likely few other example in the world of a central and planned use of scientific parks which have been so quickly integrated within the city, shaping and modelling the urban environment.

Table 4 Comparison between the High Tech Development Zones (HTDZ) in Beijing, Shanghai and Chengdu

City	Type of Park	Management	Industries	Soil consumption	Sprawl
Beijing	1 HTDZ	Centralized	ICT, software (PMI)	Small	Small
Shanghai	Multizone	Per zone	Mechanic, petrolchemistry, pharmaceutical (SMEs)	High	High
Chengdu	1 HTDZ in two areas	Central	ICT, software, bio-pharmaceutical (MNCs)	Small	Small

Note: SMEs=Small and Medium Enterprises; MNCs = MultiNational Companies; ICT = Information and Communication Technologies.

REFERENCES

[1] BREZNITZ D, MURPHREE M. Run of the red queen[M]. New Haven: Yale University Press, 2011.

[2] WOETZEL J, MENDONCA L, DEVAN J, et al. Preparing for China's urban billion[J/OL]. McKinsey Global Institute, 2009. http://www.mckinsey.com/insights/urbanization/preparing_for_urban_billion_in_china.

[3] MILLER T. China's urban billion[M]. London: Zed Books, 2012.

[4] SIDDIVÒM. Cluster industriali: il primo passo verso la endogenous innovation[J]. Italia: Mondo Cinese, 2013, 150:86-89.

Ports and Cities in the Age of Box Capitalism

Marco d'Eramo
Italian journalist and writer
marcoderamo@ hotmail. com

ABSTRACT: In this article I will examine the urban and urbanistic effects of the "containers' revolution" in the maritime transportation. For this purpose I will distinguish three main stages of the maritime technologies adopted by the capitalist entrepreneurs: the sailing capitalism (from Middle Age through the 18th century), the steam (and motor) navigation capitalism (the 19th century and first half of the 20th century), and the containers (or box) capitalism (since the 1960s).

I will consider the relationship between seaside cities (and metropolis) and their harbour in each of these three stages. In the age of sailing capitalism, the harbour was at the hearth (spatially and socially speaking) of the city; during the steam (and motor) capitalism age, the city turned its back to the harbour moving it at its edge, as the port became more and more and industrial plant. Eventually, with the containers' revolution, the city broke its millennial relationship with its harbour so that we have now, for the first time in history, seaside cities without ports and ports without cities.

Key Words: Amsterdam, Rotterdam, port, sailing capitalism, coke capitalism, boxed capitalism

Amsterdam and Rotterdam, the contrast between two cities can not be bigger, even if they are just 30 miles apart and share a common secular history. Rotterdam is today's European biggest port. Amsterdam was three centuries ago world's biggest port. Amsterdam is full of smoky bars and pubs, Rotterdam is all glass, steel and cement. The 20th Century architecture has been marked by architects firms like Van den Broek & Bakema, or Brinkman & Van der Vlugt, buildings by Rem Koolhaas, or Ben van Berkel (who designed the Erasmus Bridge).

Amsterdam has been preserved for centuries, while for Rotterdam the Luftwaffe took care to give space to the most incredible architectural utopias, when on May 14 of 1940 razed the old town with a carpet bombing and reserved to Rotterdam the same destiny Americans would reserve to Dresden and Hamburg.

Amsterdam is the port of what I call the *sailing capitalism*, the capitalism which accumulated wealth by sailing in galleons and caravels through the Oceans in search of spices, slaves, cotton(Figure 1). Of this sailing capitalism, Amsterdam has been for long time the world capital. Visiting the city in 1774, the great French philosopher and novelist Denis Diderot writes that Amsterdam "is the market of the universe". The "sailing port" is a place of incredible density and crowd. And Diderot says "The port is wide a quarter a league (two third of mile) and long half a league (one mile and half). When I saw it, it was full of ships whose masts formed a forest. Nothing on the earth awakens the idea of such a portentous opulence. In this port there were two-three hundred vessels, and in winter their number is normally of roughly 1,200."

Figure 1 Hendrik Cornelisz Vroom (1562 ~ 1640), Amsterdam port

The crowd of ships corresponds to the density of humans. Port industry is labour intensive. Measured by the tonnage, the crews are incredibly large. Loading and unloading is performed manually, just as the maintenance of sails and of rigging. And the density of human workers generates a crowd of underworld, tavern servants and prostitutes who live of the port wages. The sailing port is quintessentially a picaresque port teeming with different races, resonating a babel of languages, curses, blasphemies, overflowing with exotic goods, perfumes and stenches, from the salted cod of Terranova to the Ambon's clove, to the West India tobacco. The "sailing port" is in the same time vibrant and stinking: "Amsterdam is an infected city."

The Amsterdam tree-lined channels, the cosy 17th century mansions are just the patrician face of the commoner's port. In the sailing capitalism the seaside cities rally around their port. The port is the navel of the city. Its pumping hearth. The same happens everywhere, in Amsterdam with the Damrak and the Dam, in Marseille, where the town slopes down to the Vieux Port, but even in Genoa or in Venice where the Canaletto's paintings show the masts just behind San Marco square. In these cities the patrician mansions of the rich merchants are never too far from the docks and from their vessels. Hence the quays' centrality.

The main function of quays in Dutch ports was the mooring, loading, and unloading of ocean-going ships that were, as such, part of a world-wide infrastructural network. But, at the same time, these quays were lined with housing and other urban facilities; used as urban traffic routes and promenades, Dutch quays were an inextricable part of the city's network of streets. The quay functioned as anintermediary system that blended together two different networks—international and local—which functioned on two extremely different levels of scale. The quay, as a blend of urban and international networks, was a direct spatial reflection of the blend of urban socioeconomic relations and international trade. The quays of most consequence in the shipping trade were, at the same time, the centres most vital to public city life and the places in which the most important institutions of urban administration, trade, and religion were located. This certainly applied to the principal ocean superpowers of the sixteenth and seventeen centuries: Venice, with the Riva degli Schiavoni and, next to it, Piazza San Marco; Genoa, with Piazza Caricamento; and Amsterdam, with the Damrak and, next to it, the Dam.

The second type of port corresponds to the industrial capitalism, to the "steam capitalism", or, if you like, *coke capitalism.* As the vessels discard their sails and are covered of chimneys, they change the harbours they are mooring at and transform the cities of those ports.

The same happens around the harbour that arrives around the rail station in Coketown. Coketown was the name given by Charles Dickens to his metaphorical city in the novel Hard Times, and Coketown is how Lewis Mumford describes the 19th century industrial city. Because of logistic reasons and of transport facilities, in Coketown the factories rise just near the railway (and near the rivers). In those times there is no collective transportation, so the factory workers have to live and to sleep near their factory. And the neighbourhood around the rail station becomes a neighbourhood of factories and slums. Coketown is ruled by the triptic *factory*, *railroad and slum* (Mumford), and little by little the area near the station becomes no man's land, becomes an inner city, the first landing area for immigrants who come to work in the factories.

For the same logistic reasons, the steam port, the Cokeport, becomes a rail terminal and is surrounded by factories. The night reigns on Cokeport, like on Coketown, and black is its dominant colour. Black clouds of smoke come out of chimneys and the railroad tracks sow everywhere slag and soot. That's why the shippers, the ship owners', the factories' owners move their cosy mansions away from the port and from its smoke, and little by little the *steam city* turns its back to its steam port, to the waves of immigrants who land there and there are abandoned by the undertow. If the sail port was picaresque, the steam port is indelibly proletarian. And then the size of ships grows enormously, they draw much deeper, so the ancient ports become too small, their waters are too shallow. And so the port moves from the centre of the city, or the city centre moves away from the port.

In New York, for instance, even in 1956, long after WW II, at least 90,000 manufactory jobs were directly linked to the port activity. And if you counted all other linked jobs, shipyards, dockers, lawyers, accountants, insurers and so on, *the live hoods of half a million workers may have depended directly on the port.*

The new port was surrounded not only by industries, but it organized itself as an industry, even if a peculiar one. The metallic hulls permitted new incredible tonnages, the motor engine accelerated the speed of navigation, so the world trade boomed. And the volume of merchandises to be loaded and unloaded in the ports escalated exponentially. So the number of dockers grew to reach the size of blue collars teams in the factories. To get an idea, in 1951 more than 51,000 longshoremen worked in the New York Ports, while in London they were 50,000. That's why over all the world great Unions of longshoremen established themselves: remember the epic strike of the longshoremen on the East coast of the Us in the 1930s'. These unions controlled all the harbour, gently or not, abiding the law or not, as you can see in a fictional way—but not too far from reality—in the Elia Kazan's movie (1954) *On the Waterfront* starring Marlon Brando.

One could say that in the 50s' of the last century, the port was certainly an industry, but not a Taylorist one, it was not organized according to the principles of Fordism. There was not any assembly line. A loading or unloading line was impossible because of the merchandises' irreducible heterogeneity. Sure, there were mechanical elevators and conveyor belts, but the most of the work was still performed by hand. Relatively small ships carried incredible varieties of goods: bananas' bunches, cements bags, whisky barrels, cotton bales, copper bars. Each of these goods had to be unloaded separately from trains or tracks, placed on a pallet prior to storage. Then it had to be inventoried, and then put on the dock, and on the dock counted again and then loaded piece by piece on the vessel in a way to maintain the centre of gravity low, and to balance the weight in order to avoid any capsizing. At the same time, the order of the loading had to be the inverse of the unloading: the last goods to be stowed had to be the first to be unloaded in the first call port.

It was a long, long job: in New York the time needed to load the bulk was roughly 2 man hours per ton. A cargo, which carried 5,000 tons of merchandises, needed 10,000 man/hours, i. e. 12 days eight hours each for five crew of 20 men. At every call the ship was berthed at the pier for weeks and weeks, and spent much more time at the dock than at the sea.

The final result of all these factors was that 60% to 75% of the total cost of shipping was generated by the vessel berthed at the pier. And it was exactly because the sheer size of this cost that factories and plants were located near the ports, just to minimize the expenditure. For the same reasons, industries found convenient to produce goods in proximity of selling points. It was uneconomic to buy in Europe a TV set produced in Asia since the cost of shipping (included the loading and unloading) was such a big part of the final price. No matter how cheap was the labour in Asia or in Mexico, any goods produced far away were more expensive than goods produced in the consumers' area.

Therefore shippers, ship owners, export industries, all were looking to a way to solve two problems: (1) how to neutralize the excessive power of the Longshoremen's Unions, (2) how to reduce shipping costs, and particularly port operations related costs. And they found a solution.

We are now back to Rotterdam with its huge steel and glass buildings, large windswept avenues, and with its port, the biggest in Europe. This port corresponds to the third age of capitalism, an age I will call *boxed capitalism*, or, if you like *canned capitalism* (in the way you say canned meat). An age that started roughly in the 1970s' of the last century. After the *merchant capitalism* driven by the different India Companies, after the *industrial capitalism* driven by the German Krupp, the American Ford and General Motors, now we live in the age of the *distributive capitalism*, whose the leading figure is no longer the merchant or the industrialist, but the distributor, and whose the true, silent protagonist is the consumer.

The new economic leading figure is Walmart, the colossal chain of low price goods. Low price because these goods are bought where they cost less, where the labour cost less, in China, Mexico, Laos, and then sold at home exploiting the work of underpaid salesmen, of illegal cleaning workers, prohibiting any form of unionizing.

As the legend goes, the container revolution was set off in 1957 by a genial North Carolina's hauler, Malcolm Mclean. Actually, as it happens very often with the most important innovations and inventions, the containerization had many fathers, and probably many mothers. Many people had already tried to reduce costs and to weaken Unions by shipping their items pre-stored in a metallic boxes, packed directly in the factory, loaded on a truck or on a train and then directly in the vessel's bulk, bypassing intermediate inventorying, storing and, most important, bypassing the longshoremen crews supplanted by cranes and bridge-cranes. The time at the pier would be drastically reduced, so the ship would carry out more crossings in the same time span, generating more income and reducing costs. The sealed box would avoid the constant, epidemic thefts that characterized the port activities. Less thefts meant less insurance premiums. With the container it doesn't matter if the items travel by ship, by train, or by road, and pass from one support to the other without any further intervention: transportation becomes truly inter-modal.

One could suppose that container was the "Columbus Egg", better the "Columbus Box". But it was not. Container would never become what it has become without another decisive element, the *standardization*. We never notice the sheer power of the standardization, the telephone plugs that work only because they are all the same, screws and bolts we can buy indifferently only because the pitch and the calibre are always the same everywhere. Had the containers not had common standards, every container could be used only by a single shipping company, for a single kind of truck or railroad car and could be lift only by a particular crane. More, different containers could not be easily piled on each other or beside the other, with enormous problems of stability for the ships. Without standardization, the containers could not have been interchangeable.

The revolution was won not by the container, but by the standardized container.

And to standardize is not an easy task, and not always succeeds, for it runs into the force of the existing. Four centuries after the birth of the new physics, with Galileo Galilei, the world industry has not yet agreed on a common standard of lengths, because yards, feet and inches cannot be easily converted in meters and centimetres. In Europe, after more than a century of telephones and electricity, we still have different plugs and sockets in different countries. In America cellphones work on a different frequency than in Europe or in Asia.

Standardization is always met with strong resistance. And container standardization met extraordinary resistance. Each carrier wanted to maximize the capacity of his own vehicles. The American haulers wanted containers exactly as large as their trucks. Rail companies wanted containers exactly as large as their rail cars. For some goods too heavy, shippers wanted smaller containers, in order to be lifted by less powerful cranes.

It is hence extraordinary how short time the process of standardization of containers did take to succeed. In 1961, only four years after the conventional start of the Containers age, all the different actors agreed on a common standard: and the agreement was made possible only by the need of getting rid of the Longshoremen Unions. The agreement established that each container has to be 20 foot long, 8 foot wide and 8.5 foot high. Then there is a possible *double container*, 40 foot long, same width, same height (the 20 foot containers actually measure 19 feet and 10.5 inches, to make easier to store two single containers in the space of one double). They agreed also on a particular form of coupling and jointing the containers each other (in order to pile them on the ships, without moving in case of rough seas).

But even with this fast agreement the new technology would not have won without a factor that shaped the most modern innovations: the war. In fact it was the Vietnam War between 1965 and 1975 that delivered the final victory to the containers' revolution, with the need of supplying an expeditionary force of more than 500,000 men across the Pacific Ocean.

But the Vietnam War pushed towards the globalization by another mean: the container ships came back empty from Vietnam, with the Pentagon paying the two ways trip. So it seemed even too natural to make a call in Japan in the way back, and carry Japanese goods to the US: the shippers could offer discounted fares for shipping these goods since the trip was already been paid by the American military. It was in this way that Japanese television sets, recorders, appliances and cars invaded the United States.

The immediate goals of containerization were reached beyond any hope. The dockers' number collapsed everywhere in the world, even if the volume of the international shipping is now more than ten times what it used to be before the containerization. Consequently the power of the Unions was significantly eroded. The delay between sending and delivery was cut spectacularly by the vertical fall of loading and unloading time: in the 1950s' 2 man hours were needed to stow a single ton in the New York's docks; now a container of ten tons is stowed in a couple of minutes by a single operator of a bridge crane. So the container takes a decisive step towards this "annihilation of space by time" Marx considered the fundamental characteristics of the capitalist exchange.

But the container generated cascading effects. "The container is much more than a box, it is a vector of production and distribution."

Certainly, nobody predicted that intermodal transportation would destroy millions of jobs in the industrial world and would resurrect them thousands of miles away. And this happened because it was no longer necessary to produce goods in the vicinity of the selling point. That's why the half million jobs linked to the port of New York melted in the air, and New York ceased to be an industrial city, and relied only on the services sector. The measure of this revolution in the geographical distribution of the manufacturing centres can be read in the paragon between the twenty top-ranking container ports of the world, ranked by the number of million of standard containers units (TEU: Twenty-foot-equivalent-unit, the space of a 20x8x8,5 foot container) that transit every year.

We are struck by the huge growth of the containers' global volume. In 1989 the total sum of containers in transit in the 20 top ports was of 44.5 million twenty-foot equivalents units. In 2009 to get beyond this threshold, the first two ports were enough. Singapore and Shanghai alone made 50.8 millions TEU. In 2009, 210 million TEUs passed through the 20 top ports, as many as five times the amount of 20 years before. And in 2012, the traffic through the top 20 ports had grown of 30% over the past three years and was almost seven times the traffic of 1989 (Table 1).

Table 1 The 20 top containers ports in the world (in millions of TEUs in transit)

	2012		2009		1989	
1	Shanghai, China	32.5	Singapore	25.8	HongKong	4.5
2	Singapore	31.6	Shanghai	25.0	Singapore	4.4
3	Hong Kong, China	23.1	Hong Kong	20.9	Rotterdam	3.9
4	Shenzhen, China	22.9	Shenzhen	18.2	Kaohsiung	3.4
5	Busan, South Corea	17.0	Busan	11.9	Kobe, Japan	2.5
6	Ningbo-Zhoushan, China	16.8	Guangzhou	11.2	Busan	2.2
7	Guangzhou, China	14.7	Jebel Ali, Dubai	11.1	Los Angeles	2.1
8	Qingdao, China	14.5	Ningbo	10.5	New York New Jersey(25)	2.0
9	Jebel Ali, Dubai	13.3	Qingdao	10.2	Keelung, Taiwan, China	1.8
10	Tianjin, China	12.3	Rotterdam	9.7	Hamburg	1.7
11	Rotterdam, Netherlands	11.9	Tianjin	8.7	Long Beach, USA(23)	1.5
12	Port Kelang, Malaysia	10.0	Kaohsiung	8.5	Yokohama, Japan	1.5
13	Kaohsiung, Taiwan, China	9.8	Antwerp	7.3	Antwerp	1.5
14	Hamburg, Germany	8.9	Port Kelang, Malaysia	7.3	Kehin	1.4
15	Antwerp, Belgium	8.6	Hamburg	7.0	Felixstowe, Britain(36)	1.4
16	Los Angeles USA	8.1	Los Angeles	6.7	San Juan, Puerto Rico	1.3
17	Dalian, China	8.1	Tanjung Pelepas	6.0	Bremen, Germany(21)	1.2
18	Keihin, Japan	7.8	Long Beach	5.0	Oakland, USA	1.1
19	Tanjung Pelepas, Malaysia	7.7	Xiamen	4.6	Seattle, USA	1.0
20	Xiamen, China	7.2	Laem Chabang, Thailand (24)	4.6	Manila, Philippines(38)	0.9
	Total 20 ports	286.8	Total 20 ports	210.0	Total 20 ports	44.5

But there is more. In 1989, among the 20 top ports, five were Americans, five Europeans, two Japanese, two of Taiwan China, and there was not a single Chinese port (at that time Hong Kong was still under British rule). 23 years later, in 2012 the rating has turned upside down: to find the first European port you have to go down to 11th position

(Rotterdam which was the third world's port); only one American port remains among the top 20 (and it is in 16th position). Long Beach and New York/New Jersey are out of the top 20 (the first in position 23, the second in 25), while Oakland and Seattle out of the top 50. Manila is out of the top 20; Japan and Taiwan China have just one port retained. But today 9 over twenty ports are Chinese and they move 151 million TEUs, more than half the total.

Relocation has not been caused by containers, but without them it would never have reached the titanic dimension it has. And it is a self-feeding process. Actually someone has to produce these hundreds of millions of containers travelling around the world. And who makes it if not China which produces 93% of circulating containers?

But asymmetry has another unexpected consequence. Every 100 containers arriving in Europe from China, 40 come back empty. For Russia the rate is 70%. The problem of empty containers worries the shippers and inflates the statistics (they don't distinguish between empty and full containers), causes an environmental problem and push up the shipping costs: in 2004 moving empty containers had a global cost of 5 billion dollars a year. And in crisis times empty containers pile up in the ports. In March 2009 in the port of Busan (fifth port of the world), 31,700 empty containers cluttered all the piers. The emptiness of the box is the nightmare of the shipper.

But there are other unintended consequences. The fact is that the containers obey to the logic of fixed costs. From this point of view, its logic is similar to that of the 19th century railways or todays airlines. An American airline has to buy its planes, to build its terminals, its maintenance hangars, to pay the crews, and so on, no matter if the planes fly empty or full. In the same way a shipping company has to maintain ships and terminals no matter if the ships travel empty or loaded. And ships are more and more expensive to create economies of scale. Today, the biggest container ship of the world, launched in 2013, the Mærsk Mc-Kinney Mϕller, can load an astonishing amount of 18,270 containers in a single trip: lined one after one other they would form a line 70 mile, or 113 km long! Such a vessel needs special terminals, specifically equipped, and therefore further investments, further locking up of capital. Hence companies are ready to compete on the busiest routes (on which they will offer the most appealing discounts), while fares will be much higher on the marginal, peripheral route where there is no competition. It is the same reason why a ticket London New York is less expensive than a ticket Youngstown Milwaukee, or Venezia Palermo, if the distance from London to New York is five times longer. It means that those who need to send his goods to a small centre will find more convenient to sent the items first to a great port which has competitive advantages and then, from the big port, with a single short leg, to the final destination. We see that for the container works the same logic that created the hubs for the airlines: you have big sea freeways from hub port to hub port, and then secondary sea roads to small regional and local ports. Like in the air hubs, in the container hubs the traffic is essentially a transit traffic, a tranship traffic.

This logic has an urbanistic consequence that nobody foresaw: in a given region only few ports survive and hence many ports die. Until today any big seaside city had necessarily a port, and any big port was necessarily in a big seaside city. Now this is no longer true. For the first time in the history you have big sea cities that have no significant port: it happened to New York, San Francisco, Liverpool. Sea cities without ports, this is another reality generated by the containers.

The container transforms the very nature of ports. First, the port has been privatized. In the past, the port was the aquatic piazza, the marine or fluvial equivalent of the forum, of the agora, in the same time market and public place of meeting. It was unthinkable that the port was not ruled by the government of the city or that of the nation. But now the divorce between city and port, and the sheer size of the investments needed to handle such huge loads, makes the port a business like any other. More: the category itself of port has become useless from the heuristic point of view, and has been replaced by the category of terminal: it has been proved that the traffic of any single terminal of a port is unrelated with the traffic of the next terminal. Notion itself of port disaggregates, fragments itself.

More, the container changes the physicality of the port. Since the containers do not need any more to be stored, they have to beparked, then the containers port needs endless parking spaces, at the measure of the planetary traffic. And, after the parking, you need colossal reservoirs to store the liquid or solid load travelling bulk (oil, minerals, corn, wheat, rice, fruit juice). And then you need parking space for the trucks, and then dead rails space for the rail cars waiting to be loaded. Already in 1995, when the container ships had a capacity of only 4,000 ~ 5,000 containers, any single berth needed a surface of 16 hectares to manage its load. The container port has an insatiable voracity of spaces, wider and wider. The biggest world's port, Shanghai, covers a staggering surface of 1,414 square miles (4.580 km^2), more than half the Delaware. To the container port the city is too tight. Exactly in the same way of an international airport, a container port likes better open country or even the open sea. Shanghai's port acquired a terminal in deep waters in the East China Sea, the Yangshan Deepwater Port 19 miles from the mainland connected to it by a freeway 20 mile long.

Even without reaching such extreme solutions, you have to note that the biggest Italian container port is not Genoa or Naples, but it is a small town called Gioia Tauro, of 18,000 inhabitants, downtown nowhere, in Calabria. As container port, Gioia Tauro was born only in 1994, 17 years ago, but since many years it outpaces the two historical ports of Italy. With an annual traffic of 3.4 million TEU, Gioia Tauro is the fifth container port of Europe, after Rotterdam, Antwerp, Hamburg and Bremen.

The same happens in Great Britain where until the 1960s' Liverpool and London absorbed together the half of the maritime British traffic. Now they host a negligible share of this traffic and the main container port is Felixstowe, 90 miles northeast of London on the North Sea: in the 1960s Felixstowe was just a small village and had not even a dock.

And now, come back to Rotterdam: even if it is small compared to Shanghai and Singapore, it is still the biggest European container port and covers an area of 41 square miles. It is so huge that the silence overpowers you. Before seeing it, you imagined a lot of sounds: yells, sirens, tuds, engines. Instead you hear nothing. You turn around and it is so huge that it looks like a desert, even if it is the busiest sea call of our continent.

The sailing port was full of smells, colours, sounds, people. The container port is empty, silent, nothing smells, every merchandise is hidden from the gaze by the box, it becomes an immaterial, aseptic abstraction. In a container it could be anything: nutmegs, or cell phones, or cases of whiskey of even fresh fish in a refrigerated container: from outside everything is like everything (the indistinguishability corresponds to the new tariffs: shipping is not charged by the weight or by the kind of the shipped items, but by containers units). In the port of the *canned capitalism*, merchandises move like the money in an electronic transaction: invisible, colourless, odourless.

This invisible merchandise seems to move on its own. The ships are berthed on 25 miles of piers, so that in your visit you see only one huge ship and only once a while. And since the sea is 15 miles away from the city, you have the strange situation of a port without sea.

And what a loneliness! In fact, even if it is the busiest container terminal in Europe, 6 middle size (5,000 TEU) container ships every day would be enough to dispose of the whole load of 10 million containers. Since everything is automatic. Since three bridge cranes can unload an entire container ship, six vessels a day need only a staff of few hundreds people scattered on 25 miles of docks. Actually the port is a desert. Of course, the port of Rotterdam has many other employees and hired staff, for a total of 70,000 people, but the overwhelming majority works in the ground services, and not as dockers. In the port of Los Angeles, on 16,360 workers, 85% of them are warehousemen or teamsters. In Rotterdam the traffic is controlled by satellite and by 28 automatic radar stations. The entry of the biggest vessels has to follow the surge of the tide (the deepest dock is 25 yards deep and the biggest ships draw 24 yards).

But even this dimension is too tight now. "In Northern Europe there is the proposal for a major new container transhipment terminal at Scapa Flow, in the Orkney Islands, at the very point where the North Sea meets the Atlantic Ocean (Pentland Firth), already a long-established location for ship-to-ship oil tankers transshipment". Scapa Flow is 600 miles away from Rotterdam (and even more from Bremen, Hamburg and Antwerp), but it aims to attract most of the traffic directed right to these ports, because of the time sparing (in the transshipping operations) and the logic of the hub: a hub has not to be located near a city or on the mainland, a hub has to be convenient for the oceanic routes hub-to-hub.

CONCLUSIONS

Each stage of globalization marked a growing separation between seaside cities and their harbours. In the sailing capitalism age, the port was the city's hearth, just next to the cosy mansions of rich merchants and government buildings; in the steam capitalism age, ports became the industrial outskirts of the seaside city. In the age of canned capitalism, the sea cities, or the cities in proximity of the sea have no more relation with their ports. Some cities divorced when they lost their ports (liners are s marginal activity, today nobody really travels by ship): for example London, Liverpool, New York, San Francisco (remember that the steam capitalism had marked the end of eminent harbours, Venice first of them). Elsewhere the relationship has been loosen, and the ports are departing from the cities, as in Shanghai, or Rotterdam itself. Or you have now ports without city, ports in the middle of nowhere, as Gioia Tauro, or Felixstowe, or tomorrow Scapa Flow.

The difference between the two situations is that, even indirectly, the big abstract port brings an economical advantage to the nearby city, while the city which lost totally its port has to recycle itself. The "port orphan" cities usually recycle in the service sector. Only few of them (London, New York) drive their financial vocation, but all of them bet on tourism. No reconversion is more symbolic than the tourist recycling of docks in Piers and Waterfronts. It is a paradox that the urban section of the Rotterdam's port has been reconverted in tourist/residential Waterfront; the big shipping company Holland America's headquarters are now Hotel New York; and the disused storage Vrij has been recycled in the Exotic Festival Market. The same happened to Pier 30 in San Francisco, to gentrification of London's Docks, to the Liverpool harbour where the Albert Dock now hosts a theme park with a Ferris wheel, a Slavery museum, hotels, restaurants and a Beatles memorabilia shop. The tourist recycling of the port cities is direct consequence of containerization. Liverpool historical path is extraordinarily expressive: born as slaves port, it became the most important colonial port of the British empire, now it recycles itself as a tourist city, even if there is almost nothing worth to be seen in this city.

The lightness of the entertainment industry as direct effect of the overwhelming materialness of billions of tons moved by the world shipping. The interesting thing is that all these recycled piers and waterfronts look very like each other, their recycling is as standardized as the containerization that caused it.

The last unintended consequence is the disused container. Hundreds of millions of metallic boxes, once completed their task-usually after 6 ~ 7 years of honourable maritime career-become a big environmental problem. More and more often on every continent the most different landscapes of our planet are marked by the same identical wrecks, archaeological ruins of our modern age. Containers' recycling becomes an urgent goal. The simplest way to recycle a container is to use it to make another container. But the new trend is to use the container to build buildings.

Container home is the extreme version of the mobile home, the prefabricated home on wheels so frequent in the US. You have to note that the mobile homes are as standardized as the containers. And, of course, like the mobile home, the container home is the residential solution for the less well-off, for people border line to poverty.

In Africa, abandoned containers are particularly frequent since the ports are not equipped to move them. After have been emptied of the merchandises' load, they are sold for 7,000 rands (1,000 dollars) for a single container, a 20-foot, and little more, 8,000 rands, for a double, a 40-foot container. Khayelitsha is a black suburb of Cape Town. In Khayelitsha containers are used not only as houses, but as offices, fast-food shops, cell-phone shops, even as medical clinics (metallic walls resolve any security problem). In 2005 the tallest container building in Africa was a school dormitory, in a western province of South Africa, built using 47 containers as modular units.

But now the container is trendy for high standing buildings as well, for instance the Riverside Building in London, overlooking the Tames, built in 2005 in only 8 days assembling 72 containers to obtain 22 office units. Since 2006 a container has been used in Italy for an itinerant exhibition called "Container Art"; and in 2005 the University of California at Santa Barbara launched the "Container Project" with the participation of several artists, and in 2007 organized the forum *The Traveling Box: Containers as the Global Icon of our Era*. And are now countless the books on containers' architecture.

While breaking the relationship between port and city, the container is becoming a constituent element of the future city.

The most unintended-and ironical-consequence of containerization is the security concern. Sealed at departure, and unsealed at arrival, the container hides merchandises from the gaze, but in the same way it can hide atomic bombs, biological weapons, terrorists posses, cocaine or illegal immigrants. The sheer quantity of travelling containers makes it impossible to control them. An US coast guard commander estimated in 2004 that to thoroughly inspect a 40 food container a five man team needed 3 hours. To inspect all the containers arriving to Los Angeles/Long Beach ports you need 270,000 man/hours a day, i. e. 35,000 agents. And these are relatively small ports. And by the time, these inspections would block the turn-over of ships berthing at the pier. The world trade would be stopped. In the age of the so called "war on terror" the container hides a threat on our safety. Nobody imagined that our civilization would be obsessed by the Box Nightmare.

REFERENCES

[1] MEYER H. City and port: urban planning as cultural venture in London, Barcelona, New York and Rotterdam[M]. London: International Books, 1999.

[2] DIDEROT D. Voyage de hollande (1773), in oeuvres inédites de diderot. Le neveau de rameau. Voyage de hollande[M]. Paris: chez J. L. J. Brière, Libraire, 1821.

[3] MEYER H. City and port[M]. London: International Books, 1999.

[4]LEVINSON M. The box: how the shipping container made the world smaller and the world economy bigger[M]. Princeton: Princeton University Press, 2006.

[5]MERCOGLIANO S R. The container revolution[J]. Sea History, 2006: 8-11.

[6]MARX K. Outlines of the critique of political economy[M]. Penguin: Penguin Classics, 1993.

[7]RODRIGUEZ J P, NOTTEBOOM T. The geography of containerization: half a century of revolution, adaptation and diffusion[J]. Geojournal, 2009,74: 1.

[8]RIMMER P J, COMTOIS C. China's container-related dynamics, 1990-2005[J]. Geojournal, 2009,74: 33-50.

[9]DERAMO M. The pig and the skyscraper-Chicago: a history of our future[M]. London,New York: Verso, 2001.

[10]SLACK B. Ports, cities and global supply chains[J]. The Terminalisation of SeaPorts, 2007: 41-50.

[11]VAN DE VOORDE E. Sea ports, land use and competitiveness: how important are economic and spatial structures[C]//BANISTER D. Transport and urban development. London: E&FN Spon (Chapman & Hall),1995: 218-240.

[12]HALL P V. Container ports, local benefits and transportation worker earnings[J]. Geojournal, 2009,74:67-83.

[13]BAIRD A. The development of global container transshipment terminals[J]. Geojournal, 2009, 74:73.

[14]KOTNIK J. Container architecture[M]. Barcelona: Links Books, 2008.

[15]HERBERS J. Prefab modern[M]. New York :Harper Collins, 2004.

[16]SLAWIK H, BERGMANN J, BUCHMEIER M,et al. Container atlas: a practical guide to container architecture[M]. Berlin: Gestalten, 2010.

[17]GORDON A, BERGDOLL B, MCLEAN W F,et al. Quik build: adam kalkin's ABC of container architecture[M]. McLean: Bibliotheque,2008.

Low Carbon City Planning Strategies Based on Severe Cold Climate Regions

Hong Leng[1,a], *Cunyan Jiang*[1,b]
[1] *Harbin Institute of Technology, Department of Architecture, Harbin, Heilongjiang, China*
[a]*hitlaura@126.com*, [b]*hit_jcy@163.com*

ABSTRACT: With growing global warming, human activities under the background of rapid urbanization are the main reasons of the climate change. Low carbon city construction becomes one of the effective means of reducing climate warming. However, severe cold climate has direct effects on low carbon city construction. Compared with warm cities in south China, winter cities consume much more energy. This paper analysed the effects that severe cold climate lay on city energy consumption and carbon sink system, and then it found the relationship between low carbon target and city planning factors in winter cities. Finally it discussed low carbon city planning strategies in cold regions from six aspects in detail. These strategies would be the guide tactics of low carbon city planning of wide-ranging severe cold regions in China, for the purposes of improving the quality of human settlements, optimizing the influence of global warming, and strengthening the urban competitiveness in cold regions.

Key words: severe cold climate; low carbon; city planning strategies

1 INTRODUCTION

Global warming is a great challenge to the survival of human civilization in the history of mankind, and no country or city can manage alone under the climate change challenge. In the fourth assessment report. The IPCC explicitly indicated that global warming is beyond dispute, and human activities are the main reasons of climate change under the background of rapid urbanization. In order to reduce greenhouse gas emissions, such as CO_2, all the countries had done plenty of work, and the concept of low carbon emerges. City is the key to carbon emissions reduction, and it has become the hot topic of many research projects throughout the world.

Now, the rate of urbanization in China is about 52.57%, and China's urbanization is probably the fastest in the world, of which the CO_2 emissions intensity has already surpassed the U. S. as the world's biggest polluter. The quality of urbanization in China will increase immensely in the near future, and in order to confront this big challenge, China should take the pathway towards Low Carbon development. In the Climate Change Conference held in Copenhagen in 2009, China announced that there will be a reduction in emissions intensity of its economy of 40% ~45% by 2020 over 2005 levels as a platform for binding targets in medium and long-term plans of economic and social development. In addition, domestic scholars have done some research in the low-carbon city development path, then they put forward low carbon of basement, low carbon of structure, low carbon of mode respectively and low carbon of support, as shown in Figure 1.

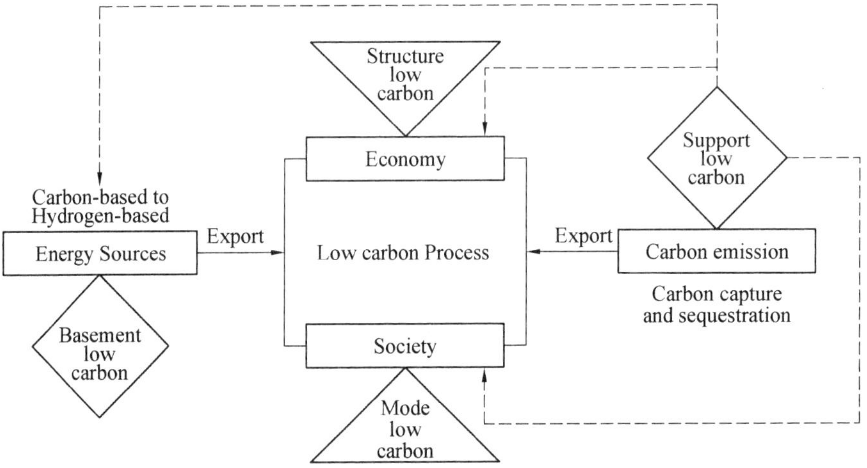

Figure 1 Framework of low carbon city development path

However, one third of total China area is under severe cold climate. There are many winter cities in China, which have cold and long winters with strong wind and snow, and it has a 6-months heating period. At the same time, due to the cold air in Siberia every year, compared with other countries at similar latitudes in the world, winter temperature of winter cities in China is far below, and the average temperature in January is always lower than −18℃. In order to overcome the effect of severe cold climate, cities need to deplete more energy, which is against low carbon city construction. City planning directly determines the scale and the distribution, lifestyle, consumption habits, transportation of urban social economic system, which is closely related to urban energy consumption and carbon emissions, so city planning is one of the most important driver of low carbon city construction. Based on severe cold climate background, using city planning methods to realize development of low carbon cities and improving the urban liveability is the key to winter cities construction in China.

2 EFFECTS OF SEVERE COLD CLIMATE ON LOW CARBON CONSTRUCTION

2.1 CITY ENERGY CONSUMPTION

2.1.1 BUILDING ENERGY CONSUMPTION

Under severe cold climate, city building energy consumption in winter mainly comes from heating. Coal and electricity energy consumption is extremely large: living space of cities in northern China accounts for 10% of whole living space in China, but the building energy consumption in northern cities accounts for more than 90%. Taking Harbin as an example, the annual heating period is 179 days, and the annual heating consumption of standard coal is 5 million tons, costing more than 2.3 billion RMB, which excludes the expense of heating for electricity and gas of some residents and enterprise units. Furthermore, wall and roof insulation of buildings increases the use of materials in winter cities, which would consume much more energy, too. A large number of coal-fired cause air pollution, which is more serious in the heating period. Taking Harbin as an example, pollution in winter is extremely serious, which accounted for 31% of the whole year, as shown in Figure 2.

Figure 2　The haze weather of Harbin in heating period

2.1.2 TRAFFIC ENERGY CONSUMPTION

City traffic energy consumption in winter cities mainly comes from ice-snow road, which leads automobile energy consumption increasing. Free-flow speed is the speed of traffic flow when the traffic density is zero, which is the reflection of operating condition supplied by road facilities. Free-flow speed of ice-snow road is 10 km/h lower than no ice-snow road, and it leads the engine idling increasing, forming the reduction of combustion efficiency and increase of gas emissions. In addition, energy consumption for warmth and power of vehicle starts increasing result in unnecessary CO_2 emissions and fuel increase of vehicle, as shown in Figure 3.

Figure 3 Ice-snow road and traffic jam after snow

On the other hand, due to the cold climate and road of ice and snow, as well as the limited road resource, real bicycle lane in winter cities is very little, which results in that people have to choose vehicle trip in winter, so the CO_2 emissions increase, the CO_2 of different modes of transport is shown in Table 1.

Table 1 CO_2 emissions in different traffic patterns (person · km)

	Walk	Bike	Motorbike	Car	Bus	Background	Train
CO_2 emission	0 g	0 g	100 g	180 g	48	61 g	179 g
Energy consumption	0 kJ	0 kJ	1 370 kJ	2 580 kJ	680 kJ	1 020 kJ	1 270 kJ

2.1.3 INDUSTRIAL ENERGY CONSUMPTION

Old industrial bases in northeast China are mainly under severe cold climate, since PRC founding, the central government has taken the northeast regions as important industrial base in China, and many of them have been key construction cities during "First five year plan" and "Second five year plan" period. However, most State-Owned Enterprises' technology development capability is weak, with backward industrial equipment, irrational industrial and economic structure, which results in large energy consumption of old industrial bases cities in cold regions, especially the coal resource consumption. Total primary energy consumption and constitute in Heilongjiang province in China are shown in Table 2. In Harbin, industrial energy consumption accounts for as much as about 60% of total energy consumption in the year 2010, raw coal consumption increases from 1.26 million tons in 2005 to 1.71 million tons in 2010, which takes a growth rate of 35.9%.

Table 2 Total primary energy consumption and constitute in Heilongjiang Province

Year	Total number (million tons of standard coal)	Percentage of energy consumption of all types account for the total number (%)				
		Coal	Fossil oil	Nat gas	Hydroelectric	Wind power
2002	6 204.2	59.9	35.1	4.0	1.0	
2004	7 515.0	65.6	30.5	3.1	0.8	
2006	7 657.3	65.2	30.0	4.1	0.7	
2008	8 347.8	68.7	26.8	4.0	0.5	
2010	9 666.8	67.4	26.2	4.1	0.9	1.4

2.2 CITY CARBON SINKS SYSTEM

Urban greening rate in winter cities is generally low. Under severe cold climate, most of the trees, shrubs and flowers are withered in winter, and only some evergreen coniferous trees are left, so photosynthesis of most plants would fail. At the same time, the capacity of carbon sequestration is weakened, as shown in Figure 4. In addition, carbon sequestration capacity is also greatly reduced because of soil freezing in winter.

Figure 4 Withered trees in winter cities

3 THE RELATIONSHIP BETWEEN LOW CARBON TARGETS AND CITY PLANNING FACTORS

3.1 LOW CARBON BASEMENT

Low carbon basement refers to the low carbonization of urban energy, in order to optimize the energy consumption structure based on coal and other fossil fuels, and in the meantime, renew clean energy should be paid attention to in winter cities. Also, besides solar energy, wind energy, geothermal energy, hydropower, biomass energy, ice and snow can be reused as the unique resources belonged to severe cold regions. The significance is not only embodied on avoiding the negative factors from the snow-accumulation grounds to the daily lives of residents, but also on replacing the consumption of traditional fossil fuels. Moreover, low carbon basement target can be realized in winter cities, with the help of reasonable city spatial form, traffic system, green space system, etc.

3.2 LOW CARBON STRUCTURE

Low carbon structure refers to the economic structure based on the low carbon industry, which determines the consumption structure of energy, even the emissions of greenhouse gases to a certain extent. Realizing that low carbon structure needs some conditions, including optimizing the layout of urban industry, upgrading the urban industrial structure and constructing the low carbon industrial system in winter cities.

3.3 LOW CARBON MODE

Human activities are closely linked with greenhouse gas emissions. Therefore the civic lifestyles have significant effects on the construction of low carbon city. In order to realize the low carbon targets, it is necessary to change the long-term lifestyle, for example, high-consumption and high-waste behaviour, with positively announcing the concept of low carbon, which mainly embodies in the transportation, housing and other aspects of life. However, civic activities are always severely restricted by the harsh climate during winter in winter cities, which lead that low carbon concept could not strictly implement. Through reasonable urban planning design of urban spatial form, land use, traffic and green space system, negative influence on low carbon mode could be reduced in winter cities.

3.4 LOW CARBON SUPPORT

Low carbon support mainly refers to the application of low carbon preparation technology. Now, some standard systems and policies about energy-saving and emission-reduction has been already published in China, for instance, the State Council issued "Chinese National Plan of Respond to Climate Change" in 2007, Chinese Academy of Social Sciences published "Evaluation Standard System of Low Carbon City" in 2008, Housing and Urban Rural Construction issued "Design Standard for Energy Efficiency of Public Buildings" in 2010, etc. In addition, there are also some macro strategic targets in related policies, for example, emissions per unit of GDP CO_2, ratio of zero carbon energy in the primary energy and so on. However, these standard systems lack the considera-tion of severe cold climate, as well as certain guidance to the construction of low carbon winter cities. Low carbon structure can be realized through the technical regulation of city management and control design based on severe cold climate.

To conclude, the construction of low carbon city aims to the low carbon basement, low carbon structure, low carbon mode and low carbon support. City planning mainly relates to the urban space organization and optimization of layout. In terms of support for the construction of low carbon cities, the relationship between low carbon targets and city planning factors in winter cities is shown in Table 3.

Table 3 Relationship between low carbon targets and city planning factors

Low carbon Targets	City planning factors
Low carbon Basement	Industrial structure, Industrial layout, Traffic network, Green space, Laws and regulations, Buildings design
Low carbon Mode	City spatial form, Land-use, Outdoor physical environment, Buildings de-sign, Traffic network, Green space
Low carbon Structure	Industrial structure, Industrial layout
Low carbon Support	Laws and regulations, Control indicators, Green space

4 LOW CARBON CITY PLANNING STRATEGIES BASED ON SEVERE COLD CLIMATE

4.1 CITY PLANNING AND DESIGN STRATEGIES

4.1.1 CITY SPATIAL FORM

Winter cities should develop relatively compact spatial form based on reasonable urban density, taking advantage of sunshine, avoiding the negative effects of wind and provi-ding suitable outdoor physical environment. With the above supports, superiorities are shown in developing public transportation. "Compact City" focuses on a variety of mu-nicipal and public facilities in city construction, so it will save lots of energy, as well as reducing pollution, preserving the bio-environment and declining the risk of traffic acci-dents in winter.

In addition, the University of Waterloo professor in Major City and Regional Planning, N. Pressman, asserts that through the process of integrating different functional districts, as well as encouraging the city space layout to satisfy any basic demand, the community will realize the self perfection of the internal functions, also in this way, some waste can be reduced, like the long-distance transportation or avoiding waste of manpower and material resources. Although these strategies are also fit for other cities, they are much more significant in winter cities, which can effectively reduce problems of transportation and energy consumption due to severe cold climate, which, in turn, contributes to the realization of the low carbon mode through the city's spatial form.

4.1.2 INDUSTRIAL SYSTEM

Firstly, structure upgrading of the original high-emission and high-pollution industrial companies in northeast industrial bases should be completed, and the high-emission and high-pollution heavy industry enterprises also should be progressively eliminated. In addition, the promotion of comprehensive technology of new energy-saving emission and the implementation of low carbon management can avert concentrated layout of the overcapacity enterprises, finally realizing the development of low carbon industrial system.

Besides, most parts of Northeast China are the major grain producing areas, which has a big advantage contributing to the primary industry. For example, Harbin relies on a variety of organic food production bases, with huge geographical advantages in food processing, transport logistics, strategic reserves and so on. In the future, the development opportunities of popularizing the green ecological agriculture should be grasped, which could lead the development of the green food processing industry, even could build the national green food centre, through enhancing the industrial guidelines and promotions.

Moreover, one of the effective ways to bring vitality and reduce carbon emissions and pollution in winter cities is to regard the winter city culture as an industry to construct, which has been confirmed by many developed countries in winter cities. Chicago, the famous "frost belt" city, was a typical industrial winter city. In 1980s, government began to pay attention to the development of cultural industry, and now, Chicago has become the second developed city in American, regardless of industry, transportation and tourism industry. In addition, the same situation happened in Sweden, Finland, Norway and other countries, in which have treated the winter tourism as one of the pillar industries, considering of the local economy. There is no doubt that vigorously developing the polar ice-snow tourism can make a great contribution to the revitalization of the economy in winter cities. Fundamentally, realizing the low carbon structure needs some conditions, including optimizing the second industrial energy structure, resolutely developing the first industry based on the promotion of natural resources and green food bases, and improving the third industry related to the characteristic tourism and cultural industry in winter.

4.1.3 TRAFFIC SYSTEM

In order to improve the urban efficiency and increase city life qualities, priority should be given to the public transportation system, which can reduce traffic energy consumption in maximum way with the characteristic consisted of rapid development, comfortableness and majority capacity. Especially, rail transportation could not only have features of large capacity, high efficiency and low energy consumption, but also provides weather protection for citizens by using the underground space within rail transportation in winter. On top of that, prerequisite for the residents should be given to use public transport that perfects transport facilities and promotes the working efficiency, including providing reservation service in public transportation for special groups, improving transfer conditions in convenient and comfortable way, creating a warm waiting environment, ensuring bus priority and so on in winter. On the other hand, special traffic technical policies should be formulated, which can reduce the negative influence on urban transportation, owing to snowfall.

In addition, in order to advocate travel by walk and bicycle when people do a short trip in winter, priority should be given to take precedence of snowmelt processing to the pavements and non motor vehicle lanes. Meanwhile, anti-slip coefficient high materials should be adapted to pavements, as well as provide weather protection facilities for pedestrians. USA Minneapolis "Skyways System" and Montreal "Underground City", for example, both ensure that citizens could access the majority of the city centres through the warm interior environments, and not have to worry about the impact of outdoor cold climate. At the same time, clean energy is necessary to the power supply devices within climate protection facilities, which can resist cold climate and reduce greenhouse gas emissions. Admittedly, energy consumption in transportation constantly increases, and even is the fastest energy consumption sector. Under this trend, there is no doubt that advocating low carbon traffic has become one of the important strategies in realizing low carbon basement and mode in winter cities.

4.1.4 CITY GREEN SPACE & CARBON SINKS SYSTEM

Under severe cold climate, consideration should be given to effects of four season's landscapes, and also to enhance the ornamental value of plants in winter, with boosting the energy of carbon sinks. First of all, establish ecological corridor alongside the predominant wind direction in the city, which aims to introduce fresh air into the inner city and could reduce the heat island effects as well. Secondly, for enhancing the carbon sequestration capacity in winter, special species should be chosen, such as elm, larch and yew, which have strong carbon sequestration ability and cold-resistant energy, In the meantime, construction of the greenbelt should be strengthened in small-scale decentralized city. Regulations in Copenhagen, Denmark, for example, require to build two new pocket parks at least once a year and to plant a large area of green climbing plants on building facades and interchange concrete guardrail, while to plant a lot of morning glory flowers in tree pools, in order to enhance carbon sequestration capacity in winter. Effects of residential greening should be sufficiently used, too. Thirdly, upgrading the urban green quantity and increasing green roofs and walls will prevent the building to acquire the solar radiation after leaves falling in winter, and the stems in the wall become a layer of thermal insulation layer again, which attached to the roofs and walls. At the same time, the green roofs and walls could adjust the indoor temperature and increase the ratio of evergreen trees.

4.2 CITY PLANNING MANAGEMENT STRATEGIES

City planning management plays a good supporting role on low carbon city construction based on severe cold climate. Some "winter city planning" should be put forward in the future city planning, especially for the unique problems existed in livable environment construction in winter cities, at the same time, the content of "winter city planning" should be reflected in the planning achievements, consisting of drawings and instructions, with the corresponding provisions in planning text.

In addition, on the base of cutting carbon and carbon sequestration targets, some special indicators which consider about severe cold climate factors should be added into universal ones, as well as some technical regulations for city planning management and control indicator system of city planning.

To conclude, comparing to the warm areas in southern, severe cold climate plays a decisive role in effects on the construction of low carbon city, through the analysis of the previous article. There is closely relationship between low carbon city construction in winter cities and urban spatial attributes, including the urban spatial form, industrial system, traffic system, green space system, etc. Besides, there are related policies of management and implements, which could support the construction of low carbon city in winter cities. The Framework of low carbon city planning strategies based on severe cold climate is shown in Figure 5.

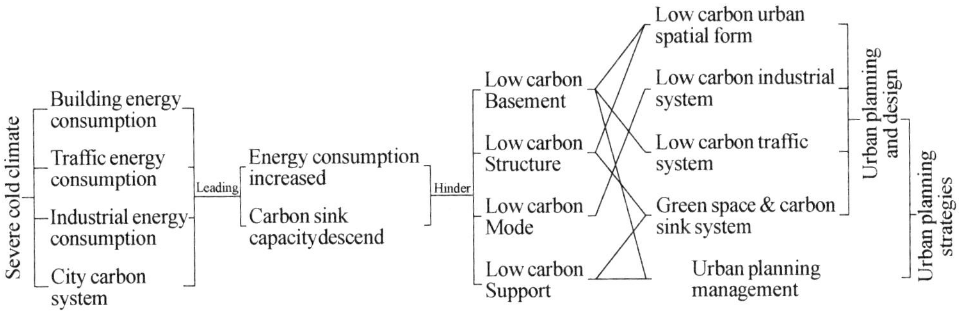

Figure 5 Framework of low carbon city planning strategies based on severe cold climate

5 CONCLUSION

China has vast land area under severe cold climate: for most areas in Heilongjiang, Jilin, Liaoning and Inner Mongolia province, the average temperature of January is lower than −18°C. According to standard of experts and scholars abroad, which says that the average temperature of January is lower than 0°C, winter cities in China will be much more. Fully paying attention to the climatic characteristics in cold regions and actively carrying out the research of corresponding low carbon city planning strategy has a extremely important significance for improving the winter city human settlements quality, optimizing the impact of climate warming, and enhancing the competitiveness of winter cities in cold regions.

ACKNOWLEDGMENTS

Supported by Program for New Century Excellent Talents in University (NCET-11-0803) and Natural Science Foundation of China (51278140).

Part 2 Enhancing the Life Quality of Rural Villages through People Empowerment

Promoting Rural Development to Combat Desertification: An Integrated Approach for Restoring Degraded Rangelands in Morocco

Laura Altea[1,a], *Marcello Lubino*[1,b]
Nucleo Ricerca sulla Desertificazione-Università degli Studi di Sassari (NRD-UNISS)
[a]*laltea@ uniss. it*; [b]*mlubino@ uniss. it*

ABSTRACT: During the United Nations Conference on Environment and Development (UNCED) in 1992, desertification, along with climate change and loss of biodiversity were identified as the greatest challenges to sustainable development.

The Nucleo Ricerca sulla Desertificazione-Università degli Studi di Sassari (NRD-UNISS) promoted several international cooperation projects in sensitive regions of central Morocco aimed at promoting strategies for rural development and combating desertification processes.

In 2003, a five year project, titled "Demonstration project on strategies to combat desertification in arid lands with direct involvement of local agro-pastoral communities in North Africa", was launched in the rural community of Ouled Dlim (Wilaya of Marrakech, Morocco) and financed by the European Union under SMAP Programme (Short and Medium term priority environmental Action Programme). This main intervention has been followed by other two projects financed by the Autonomous Region of Sardinia (Italy): RIPDES project (Rural development and rehabilitation of pastures in drylands and degraded areas) which last from 2009 to 2011 and PRO AGRO project (Enhancement of agro-pastoral productions in the rural community of M'Nahba (Wilaya of Marrakech-Morocco) through the rehabilitation of degraded pastures and the exchange of knowledge and know-how between stakeholders of Sardinia and Morocco) which started in October 2011 and was concluded in March 2013.

All the three mentioned projects promoted "agronomic-technical practices" combined with participatory approaches in order to enhance local communities' capacity to manage their fragile ecosystems in a sustainable way.

The present contribution is devoted to illustrate the approach and the main results obtained through the above-mentioned projects.

Key Words: desertification, soil erosion, Atriplex nummularia, sustainable management of land, rural development, participatory approach, capacity building

1 INTRODUCTION

Land Degradation and Desertification

The United Nations Conference to Combat Desertification (UNCCD, 1994) formally defines desertification as "land degradation in arid, semi arid and dry sub-humid areas resulting from various factors, including climatic variations and human activities".

The same definition was adopted by the Millennium Ecosystem Assessment (2005) which clarifies that "land degradation is in turn defined as the reduction or loss of biological or economic productivity of drylands".

Today, dry lands represents 40% of global land area and they are home to more than 2 billions people, 90% of which living in developing countries (Millennium Ecosystem Assessment, 2005).

More than half of these 2 billions people lives in rural areas and thus directly relies on the land for their livelihoods. Their fate depends on the availability of ecosystem goods and services viarainfed or irrigated farming, harvesting native plant products, hunting, and pastoralism. A combination of environmental and socio economic factors such as climate variability, unsustainable human activities, limited resources and increasing population size increases the pressure on limited dry land resources, which can in turn lead to rapid land degradation, threatening people's livelihoods (Reynolds, 2013).

It is estimated that approximately 6 ~ 12 millions km^2 (around 20%) of all dry lands, mostly located in Asia and Africa, suffer from serious land degradation and desertification problems (Millennium Ecosystem Assessment, 2005). In Africa, there are countries such as Morocco in which 87% of the national territory is vulnerable to desertification. This contribution presents some interventions implemented in Morocco which can be considered as a representative case study for combating land degradation and desertification.

2 METHODOLOGY

Since 2003, the Nucleo Ricerca sulla Desertificazione-Università degli Studi di Sassari (NRD-UNISS), promoted several international cooperation projects in the region of Marrakech-Tensift-El Haouz (Morocco), characterised by rural poverty, food dependency and land degradation.

The common denominators of these interventions are the overall objectives and the adopted approach and methodology, as well as the Moroccan targeted regions.

In particular, all the projects aimed at promoting rural development, enhancing the social and economical conditions of local populations, by implementing desertification mitigation strategies in highly degraded rangelands. In particular, Atriplex nummularia plantations, an Australian fodder shrub, was planted in the project area in order to rehabilitate land productivity.

Considering that local communities have a greater stake than anyone else in managing and improving their agricultural production system and at the same time they are the main responsible for ensuring a sustainable and rational management of their fragile lands (UNCCD, 1998), the interventions adopted a participatory approach and promoted a cooperative action among all parties.

Local knowledge and management systems have been used to understand the nature and dynamics of problems, local actors have been entailed in the identification of needs and in the designing of the interventions through both the implementation of participatory research techniques and the involvement of the community as partner of the Action (interactive participation: Pretty, 1995).

The active participation of local stakeholders in all the phases of the projects (diagnosis, evaluation, design and implementation) has been essential to empower them and ensuring the sustainability of the projects' actions in the long term.

Finally, all the projects carried out in the area of intervention were characterized by a multidisciplinary approach, with a project consortium composed by different Institutions, expertise and multidisciplinary teams constituted by agronomists, development experts, zoologists, agricultural extensions services, rural community representatives, etc.

3 THE AREA OF INTERVENTIONS

The projects' intervention area is located in the rural municipalities of Ouled Dlim and M'Nhaba, 44 km North West of Marrakech city.

The climate of the area is semi arid, with long warm dry season (April to October). The annual rainfall is scarce, with an average of 150 ~ 200 mm, and concentrated between November and March (Zucca et al., 2011). The landscape is characterised by a highly-degraded and fragile hilly profile, exposed to water and wind erosion which led to progressive desertification processes seriously compromising soil fertility (UNCCD, 2008).

The main productive activities are agriculture and livestock breeding. Cereals, above all barley and wheat, represent the 89% of cultivated crops and they are grown mainly as fodder for livestock (Zucca, 2011), while pastures consist mainly of sheep and goats. In particular, the breeding of meat-sheep and the production of heavy lamb are widespread in the area, due to local communities' high demand, especially during the Eid al-Adha, the Muslim festivity celebrated at the end of Ramadam (Bellal, 2006).

The land property system is for the 97% "Guich", namely the lands that were donated by the State to the local tribes as rewards for their military services and they are used mainly for extensive common grazing activities.

The population of the two rural municipalities of M'Nhaba and Ouled Dlim is 11,755 and 14,736 inhabitants respectively (RGPH marocain, 2004). The 45% of the total population is less than 20 years old and the households have an average of seven components and are based on a patriarchal system.

4 MAIN CONSTRAINTS OF THE AREA

The area of the interventions is interested by frequent and prolonged dry spells between March and September, while water resources both surface and underground are limited. The soils are calcareous and saline, with low water retention capacity and the natural vegetation cover is sparse or absent, exposing the soils to wind erosion (Zucca et al., 2011).

Traditionally the area is characterised by extensive pastoral activities, but national government attempts to sedentarize nomadic populations along with the progressive abandonment of transhumance, led to a breakdown of traditional controls on the use of grazing lands and to an increasing pressure over the local land resources.

The combination of these factors has resulted in an increasing degradation of rangelands, which can grant only a very irregular forage production. Therefore, local breeders have become dependent on imported feed resources and have been obliged to purchase expensive concentrates from external markets to feed their livestock, in particular to fatten lambs before sell them on the local market.

In general, local populations' investments on land are very limited due to lack of capital and to the difficulties in accessing to credit. In addition, a major part of rural population involved in agricultural activities lack of technical knowledge in the field of sustainable management of rangelands, sheep dietary requirements, feed nutritional properties, while the lack of genetic selection and the high inbreeding level within sheep flocks progressively caused a decreasing of fertility and a low vitality of lambs (ICARDA, 2006).

With reference to the social aspects of the communities, the illiteracy rate is high (71% in Ouled Dlim Municipality and 64.3% in M'Nhaba Municipality), particularly among women. Women have generally a low level of professional knowledge and skills and do not perceive any salary for their work inside and outside the house.

The general lack of skills and opportunities in the area, urges people, above all youth, to emigrate towards cities, especially the nearby Marrakech, and also abroad, particularly to Europe, causing a gradual abandonment of rural environments (UNCCD, 2005).

5 MAIN ACTIVITIES

One of the main activities of the projects consisted in planting 2 000 ha of Atriplex nummularia, a drought resistant forage shrub, on the highly degraded extensive pasturelands, which characterized the area. This species has proven to be well adapted to arid conditions and is appreciated for its tolerance to aridity, salinity and direct grazing (Correal et al., 1990). It constitutes an optimum source of forage for the flocks, especially during the autumn and summer.

Furthermore, the grass coverage and the plant diversity was enhanced through the introduction of some native self reseeding legume forage species (Medicago scutellata, Medicago polymorpha, Medicago truncatula) between the rows of Atriplex (alley cropping system). These species locally produced at the "Centre de Production des Semences Pastorales Khemiss Mtouh (CPSP)" of El Jadida (Morocco) are known for their nitrogen fixation properties, which contributes to improve the soil fertility. Moreover, the stubbles can be grazed during the summer season, offering an additional fodder source for the flocks.

In addition to the promotion of technical and agronomical practises, the projects paid particular attention to promote capacity building, enhance local skills and improve the governance and management of land, which constitutes the main source of livelihood for these rural communities. With regard to this, several training courses addressed to rural farmers, rural youth, women, extension agents, technicians of the Provincial Department of Agriculture (DPA) of Marrakech and other officers of national Ministries and regional organizations have been organized. The themes of the training courses have been selected together with rural communities, taking into consideration their will and needs. More in detail, some training modules focused on techniques for environment preservation and fighting desertification, Atriplex nursery management and planting techniques, management of women associations, creation and management of cooperatives and associations, promotion of additional sustainable income-generating activities. Moreover, some modules on feed rationing and meat production techniques have been organized to support the provision of new selected rams, which were introduced to improve the genetics of the livestock. In this way, the fertility of the flock has been enhanced, improving meat production and its quality and positively impacting on farmers' income.

Furthermore, along the years, some exchange travels between Morocco and Sardinia (Italy) have been promoted by the projects to share successful initiatives in the field of pasture improvement, enhancement of the meat chain and development of women textile sector. This activity involved Moroccan stakeholders, small Sardinian associations and cooperatives operating in the agro-pastoral sector and was aimed at promoting the creation of network and links between the two Mediterranean regions, stimulating the know how exchange and the socio economic development of the areas.

Finally, some initiatives specifically addressed to women have been promoted, since they constitute fundamental actors of local development in the rural society. Indeed, rural women play an essential role in the life within thedouar (village): they are responsible for households work, water and fuel collection, livestock holding, food processing, children and sick caring etc. (Young, 1993).

However, because of poor rural infrastructure and services as well as certain cultural values and social norms, women's participation in employment opportunities in the area of intervention is limited and they are often kept in a state of subordination in relation to men: a great part of them is illiterate (80%) and lacks of qualifications and professional skills. Only a very small number of households can afford fees necessary to allow young girl to attend schools and professional courses: in addition, most of these institutions are located in urban centers and parents generally do not let their daughters go out alone from the douar. In order to try to address these problems, the projects promoted literacy classes and basic accounting courses within the douar and equipped a female laboratory with looms, sewing machines, etc. Furthermore, three women associations have been formed and several training sessions on wool processing, embroidery and sewing have been addressed to their members to create income-generating activities totally managed by rural women.

6 RESULTS AND DISCUSSION

Direct field interventions, through the plantation of 2,000 hectares of Atriplex nummularia, have given a response to local breeders concerns: green fodder reserve for the dry season and for severe and prolonged drought periods has been ensured while environmental protection has been enhanced through the regeneration of land vegetation cover, resulting in an evident reduction of wind and water erosion and soil degradation.

Moreover, the gradual reappearance of some wildlife species such as partridges, reptiles, etc. in the projects' areas shows signs of biodiversity reconstitution.

From an economic point of view, thanks to the increase in the availability of forage resources, local farmer's costs for feeding the flocks have been drastically reduced. Indeed, before the projects, local farmers were highly dependent on imported feed and expensive concentrates purchased from the market, due to the scarce productivity of their pasturelands. A local breeder interviewed in 2013, said "before the project implementation I gave three feed rations per day per sheep, now with Atriplex I only give a single supplementary ration (grain, barley, beet) in the evening".

Since the beginning of the Actions, local communities' responsibility in the sustainable management of their resources, in particular rangelands, has been significantly enhanced as well as their capacity of self-organization and mobilization. Several new breeder cooperatives have been created in the area and they have received technical training. At present, eight of them are in charge of managing Atriplex pastureland. Moreover, two new women associations have been created in the two communities, supporting local women self-promotion and income generation activities. Further investigations and data collection, are needed to determine and quantify the socio-economic and indirect impacts of the Action such as the reduction of rural exodus, the creation of new income generating activities and new opportunities for young and women, the improvements in the meat chain, etc.

Through the projects' implementation it was possible to share around Atriplex plantations a new system of interest with local stakeholders and develop social learning process (Steyaert, Jiggings, 2007). The full involvement of local communities in planning and decision making and the space they had to express their changing needs and expectations contributed significantly to a successful implementation of the action and represented a point of force for its sustainability.

In addition to the technical and agronomical practises promoted in the area, the projects trigged local governance and responsibility mechanisms as well as cooperation between all parties for the management of fragile ecosystems. In this way, the approach promoted by the Actions represents a concrete and successful solution which can contribute both to the preservation of lands and rural environment and to the improvement and enhancement of communities' livelihood, key elements for a sustainable development.

ACKNOWLEDGMENTS

EU-MEDA SMAP II project "Demonstration project on strategies to combat desertification in arid lands with direct involvement of local agro-pastoral communities in North Africa", coordinated by NRD-UNISS, contract number ME8/B7-4100/2001/0132/SMAP-5.

Project RIPDES "Rural development and rehabilitation of pastures in drylands and degraded areas", financed by Regione Autonoma della Sardegna, L. R. 19/1996, and coordinated by NRD-UNISS.

Project PRO AGRO "Enhancement of agro-pastoral productions in the rural community of M'Nahba (Wilaya of Marrakech-Morocco) through the rehabilitation of degraded pastures and the exchange of knowledge and know-how between stakeholders of Sardinia and Morocco", financed by Regione Autonoma della Sardegna, L. R. 19/1996, coordinated by CNR ISPAAM, partner NRD-UNISS.

REFERENCES

[1]CORREAL E, BELMONTE C, OTAL J. Utilization of sheep on oldman saltbush (atriplex nummularia): palatability, browse, efficiency, voluntary intake and chemical composition[C]. Proceedings of the VI meeting of FAO-European sub network on Mediterranean pastures and fodder crops. IAM, Bari, Italy: 1990, 165-168.

[2]LAI A, NUTT B. Twin sowing and summer sowing: alternative techniques to introduce legume into pastures[J]. Options Méditerranéennes, 2010,Séries A(92): 97-100.

[3]LUBINO M, MORTAJI M, PORQUEDDU C, et al. Enhancement of agro-pastoral productions in the rural community of M'Nahba (Wilaya of Marrakech - Morocco) in options méditerranéennes: new approaches for grassland research in a context of climate and socio economic changes[J]. Mediterranean Seminars, 2012, Series A(12): 497-501.

[4]PRETTY J N. Participatory learning for sustainable agriculture[J]. World Development,1995, 3 (8):1247-1263.

[5]REYNOLDS J F. Desertification in encyclopedia of biodiversity[M]. Durham, NC, USA:Duke University,2013.

[6]STEYAERT P, JIGGINGS J. Governance of complex environmental situations through social learning: a synthesis of SLIM's lessons for research, policy and practice[J]. Environmental Science and Policy,2007,10: 575-586.

[7]YOUNG K. Planning development with women: making a world of difference[M]. UK:Macmillan Press Ltd,1993.

[8]ZUCCA C, LUBINO M, PREVITALI F, et al. The euro mediterranean partnership: a participatory demonstration project to fight desertification in Morocco and Tunisia. [C]// ZDRULI P, TRISORIO L G. Determining an income product generating approach for soil conservation management. Bari :IAM,2005: 317-325.

Digital Hamlets: Innovative Methods and E-Participation Tools Supporting Self-Sustaining Policymaking at the Local Level

Alessandro Plaisant[1,a], *Maria Teresa Pinna*, *Valentina Talu* [1,b]
[1]*Department of Architecture, Planning & Design University of Sassari (Italy)*
[a] *plaisant@uniss.it*; [b] *vtalu@uniss.it*

ABSTRACT: E-participation in decision making is gaining ground. An increasing number of administrations recently tries to foster innovation processes at all levels of government, supported by EU cooperation strategies and funding. The task of the paper is to test the usefulness and opportunities coming from a mix of e-participation and "proactive" decision support tools, especially in low density residential areas and districts, affected by structural limitations in terms of accessibility, loss of population, and public services consequently. Here, participation can play an important role in the user-centred design of public spaces.

The authors present an overview of the work-in-progress on an experience of policymaking in a low density residential rural area, in order to ensure self-sustainability through a series of methods and tools to encourage participation and identifying consistent and low cost policy choices for administrations. We conclude that, far from the technology-based smart city paradigm, this bottom-up approach to innovative practices for e-participation is extremely useful since it provides an administration with a simple, immediate and free use of a pool of ideas and proposals previously unimaginable. On the other hand, e-participation tools can support administrations to better and cheaply organize a number of basic services and facilities in rural areas and foster citizens and visitors to define new input and projects for small communities.

Key Words: e-participation, strategic urban planning , decision support, self-sustainability

1 CONSTRUCTING A SHARED FRAMEWORK FOR POLICYMAKING

The search of cooperation and legitimation plays a crucial role in managing common good and leads to a new attention towards decisional processes in planning. There is a vast literature about how socially sensitive planning ought to be done, but in the Mediterranean area, especially in Italy, it is hard to find trials that have been triggered for proven practices of participatory democracy, where subjects are primarily not only institutions. Although we can speak legitimately about participation of local authorities, from the municipalities to the EU through the regions of national States, this is always an inter-institutional participation, that is something different from the creation of self-organized civic networks of citizens. In Italy, where procedures are standardized and often very close, local public administrations can become the place where to develop and to legitimise new forms of above-local practices for governance, establishing a real process of empowerment.

1.1 NEW SHAPES AND PARADIGMS FOR INTERACTION: TOWARDS OPEN GOVERNMENT AND WIKI POLICYMAKING

Better organizational planning needs new forms, less top down, more participated, more flexible, more reversible, more environmentally aware. As President Barroso said in 2010, "The crisis has also exposed fundamental weaknesses and unsustainable trends that we cannot ignore any longer" [1].

In the current situation, it is necessary to clarify some elements of renewal, starting from the point that the implementation of strategic policymaking can no longer be developed only to one level of government (State, Provincial or Municipal), as in former times. Furthermore, we need to explore innovative itineraries for an interactive policymaking, open to different perspectives, supported by pluralism-oriented instruments and tools. This approach can be declined at least according to three perspectives. Firstly, rethinking the role of public action in urban governance and involving the citizens through public participation, to rule processes at all the scales in a process of sensitivity and mutual learning. That is to say, to adjust tasks and competencies of the techno-structures of public administrations to new generations tools and strategies, to deal with complex situations and to stimulate decision-making. Secondly, strategic policymaking can take root only if there is a self-organizing capacity of the territories, supported by the institutional representation, that should foster the conditions for the development of these organizational skills, where weak or even absent. For example, the future role of the conventional structures for the implementation of policies could be different, and it could be assumed by unforeseen subjects, with new forms of structuring, which may be different for different policies or projects, and, in some cases, they may consolidate on an institu-

---

1 Opening remarks of Manuel Barroso at the Press Conference on Europe 2020 (Brussels, 3 March 2010).

tional basis. Last but not least, experimenting alternative approaches in building plans, policies and projects through opportunities coming from e-participation and "proactive" decision support tools to encourage participation and identifying consistent and low cost policy choices for administrations, as well as support administrations to better and cheaply organize a number of basic services and facilities.

Note that since internet has become an active media, as a place of content production, rather than the last century mass media of passive use, people is more and more focused on the interaction and active production of contents, including co-production as a leisure (Shirky, 2008, 2010; Verona et al., 2012). Furthermore, the presence of enabling tools and easy access applications, with easy access to broadband have increased the number of daily active users on the Internet. This cultural revolution is important at least from two points of view: firstly, from the qualitative point of view, it increases the people with new opportunities to choose and to decide, in addition to affecting communication (The New York Times or Twitter), science (peer-to-peer research), work and innovation (models of social innovation); secondly, quantitatively, it is a strong opportunity to convey the ocean of free time hours, that we spend, for example, watching TV (the so-called Shirky's "cognitive surplus", 2010), to create Ushaidi or Wikipedia-size collectively edited projects. The so-called crowdsourcing, or the participation of the crowds to the achievement of common goals through the Internet is becoming a more and more known opportunity and considered by companies and by research centres (Ibid, 2012). Within this paradigm, public administration are increasingly called into question: the Open-Government (open gov) Initiative, introduced by Obama[2] in 2009, strength every agency across US federal government toward transparency, participation and collaboration by means of the internet and, in the same time, foster democratic participation in public life, by asking the public for ideas and suggestions. The prerequisite of transparency is intended as the possibility to access, analyse, monitor (and eventually reuse) data relating to the work of public administrations by citizens. It, therefore, creates the conditions to make an independent and critical information accessible to citizens and allows them to exert a conscious participation in the political life of their own country or city. Participation of citizens in decisions is an added value, capable of guiding choices in a more knowledgeable, informed and-according to the paradigm of open government-better manner, because it would activate the so-called "collective intelli-

2 "Knowledge is widely dispersed in society, and public officials benefit from having access to that dispersed knowledge" (Obama, 2009).

gence" capable of identifying problems and solutions in complex choices (Engelbart and English, 1968; Lévy, 1994; Ibid, 2012).

The Web 2.0 environment has enabled evident advantages in public interaction. It opens new forms for participatory governance that we can summarize in the concept of online or wiki-policymaking. With the start of these mechanisms " public administrations lose, in part, the control of their actions, but they get in exchange analysis capacity and decision-making by collective intelligence. And their action becomes more effective" (Cottica, 2010). We think that open government and wiki policymaking should work in parallel with more traditional strategies for policymaking and don't replace them entirely: various group-management techniques can operate together with computerized tools to represent knowledge and better explore interactively the space of opportunities. Our emphasis is not so much on informational support or on methodological aspects characterizing all the steps of the plan, as it is on developing a mix of techniques, tools and services more or less innovative, measured directly in the contexts.

2 THE SELF-SUSTAINABLE TERRITORIAL MANAGEMENT: AN INTERACTIVE AND PARTICIPATORY POLICYMAKING PROCESS

We propose an interactive and participatory policymaking process, self-sustainability-oriented, to create an interactive dimension, with the dual intent of feeding the territorial debate and exercising urban functions in a low density residential rural area.

Various levels of government, different parts of the territory and different territorial subjects join together to harmonise rules for planning and management of the forms and processes that concern fields of collective discussion, such as the environment, services, leisure, etc., depending on their role, relevance, competences and the activities they carry out, and according to their own point of view or the interests they represent. In this form of planning the value represented is not that of the single groups, single hamlets or boards, but that of efficacy of public policies on problems of collective interest and acknowledgement of the strategies and procedures agreed both in municipal and supra-municipal urban planning instruments and in regulation instruments of the boards or groups involved in the agreement.

2.1 THE PILLARS FOR SELF-SUSTAINABILITY

Self-sustainability means a shared organization and participatory management among all the actors involved. It refers not only to environmental values, but also economic, institutional, social ones, including places of residence, work, consumption. Three pillars support self-sustainability: the urban quality of life; a social interaction and new forms of cooperation through an interactive space to exchange information, people and services; a territorial equity among populations living in different places and contexts.

There is a huge literature on indicators that may be used from different spaces to measure the quality of life, but according to the Sen's (1993) Capability Approach, individual wellbeing has not to be defined only in terms of resources and utilities, but it should also be conceived and measured directly in terms of functioning and capabilities. Therefore, we refer to an effective opportunity for each community member to use places, "what they are effectively able to be and do (capabilities), given both their personal characteristics and their surrounding environment" (Blecic et al., 2013), rather than only on the basis of intrinsic characteristics. In fact, ordinary planning tools measure the quality of life by calculating the size of the elements in terms of availability of space, services, information and opportunities, without considering neither the actual use, nor the spatial distribution, which is essential to represent processes of interactions between people and their surrounding environment. In other words, measuring the urban quality of life allows to reveal the elements that hinder or limit the possibilities of use of urban services and public spaces for different city users. Nevertheless, the possibility to pinpoint the information on free editable maps, by experiences of the users, allows policymakers and urban planners to design forms, functions and rhythms to ensure each place a full, free and effective use for everybody.

With regard to the reticular hierarchical model of urban organisation, in which a city with its complementary region has a role in relation to a hierarchy/correspondence between cities of the same hierarchical order, in the current scenarios the change is to be hoped for from a hierarchical situation in which institutional representation governs territorial processes with a linear relationship of competences, to a new situation, fundamentally tied to a network of organisational capabilities, in which small territories self-organise to govern themselves. In this way, the perspectives of each hamlet will depend on the ability of each one to offer different opportunities, services and facilities based on their own environmental conditions, features and resources.

2.2 BUILDING AN ORGANIZATIONAL SPACE FOR INTERACTION

There is a vast literature about initiatives to be pursued (e. g. multi media-aided brainstorming [3]) to increase the range and the quality of projects that can be generated in a space of interaction. A space of interaction is configured as an open organisational space (for citizens but also visitors), in which relations, interpersonal exchanges and inter-subjectivity are condensed. The space of interaction, with the aid of instruments that favour participation in the territorial debate (e. g. telematics access and open mapping, but also instruments aiding decisions on and evaluation of policies), must guarantee that all subjective rationalities that concur altogether in constructing inter-subjective rationality, those representing institutional ownership, technical knowledge (the experts and the techno-structures) and the context (the citizens and all subjects involved), are represented and organised according to a system of common rules (Weick, 1997). On the one hand, it can act and become valid through appropriate administrative instruments that sanction their procedures and, on the other hand, it can define shared rules of self-organization, reinforcing the role of the administration for the governance of the territory. Thus, an interactive space, supported by specific decision support instruments and tools, linked to real services, can help to exercise urban functions in a shared way.

Some guidelines allow to clearly draw the outlines of the space of interaction in the case of low density residential rural areas: a continuous sensitivity-making process, together with procedures and tools sustainability-oriented (environmental, social, economic and institutional sustainability) about present and future actions; an accessibility, not only physical, but also informational and cognitive, to services and facilities for everybody; the implementation of community development projects; the identification and qualification of public spaces and disused buildings; the diversity enhancement; a synergistic co-operation between communities and administrative techno-structures; information and tourism promotion. The first step consists to identify who will participate in the space of interaction. Each level is expected to activate several tools for communication, information and participation for each phase of the process, supported by a Decision Support System (DSS). The 1st level is composed by citizens, organizers, local support group and planners. They are directly involved to foster use of e-participation tools and their involvement must be participative and formally made (newspaper's articles, public meetings, focus groups, DS and e-participation tools). The 2nd level is composed by the

3 Dickey, J. W. (1993)

municipalities, operators, associations, ancillary institutions, whose involvement in the process must be linked to the construction of agreements and joint programmes and, therefore, to create consensus on choices (newspaper's articles, questionnaires, public meetings, focus groups, DS and e-participation tools); the 3rd level is the above-local level of governance, whose involvement in the process must be reached on consultation (specific questionnaire, DS and e-participation tools); the 4th level is composed by the "wiki Community" and, even potentially, all visitors, whose involvement in the process should be reached by a multi-way approach (e-participation tools). Among them, we distinguish 4 categories of actors and stakeholders: wage earners and the techno-structures of the municipalities, which manage the DSS platform and implement the tools; volunteers, which use their free-time to develop the system; interactive users (city users, visitors and tourists), which express their ideas, opinions and evaluations and, at the bottom, passive users, which use the services without implementing them. In the space of interaction, everyone can take or work in the level she/he prefers, sharing their skills and knowledge, and possibly change over time according to their needs.

2.3 WHICH INSTRUMENT FOR POLICYMAKING? OPERATING MODES OF INTERACTION

A participatory policymaking process doesn't require high investment in terms of money and time. The self-sustainability policymaking process is composed by three main phases or steps: 1. the problem definition; 2. the discussion; 3. the implementation and evaluation(Figure 1). Each level of the process highlights a series of techniques and methods with their own specific interactive tools. The combination of techniques or their different use according to the different phases or functions, rather than the selection by existing techniques or the creation of hybrids, are particularly important for the construction of government policies. The aim is the flexibility of the process, so that we can modify the choices based on the needs of the problem addressed [4]. Note that there is a large number of e-participation tools, but for the purpose of this paper we will only point out two families of instruments: a) those involved on the conventional aspects of the ad-

4 Tacket and White (1997) argue that the combination of techniques can be done according to the paradigm of the four "M": Mix, by using different techniques (even in parts) for different times within the process; Modify, by adapting techniques based on the needs of the moment; Multiply, increasing the opportunities experimenting different techniques for the same goal; Match, by choosing the technique taking into account the preferences of the participants, and situations that occur from time to time.

ministration-enforcing its action and b) those that work on strategic aspects[5]. We define here as conventional aspects, those for which authorities are already required to operate and provide services to citizens. In this sense, several tool act to allows citizens to report the daily problems of their neighbourhood. This type of instrument has had a large following and countless variations, clones and applications in different fields. The system basically works well and is based on the principle that "many eyes see more than few" (Noveck, 2006; Verona et al., 2012). Then there are the tools that are customized to foster strategic processes, on the stage of policy making. Among them we recall platforms for collecting ideas and opinions (borrowed from the commercial logic of crowdsourcing), consultative instruments for co-drafting and suggesting changes in documents, speeches, behaviours (Ibid.).

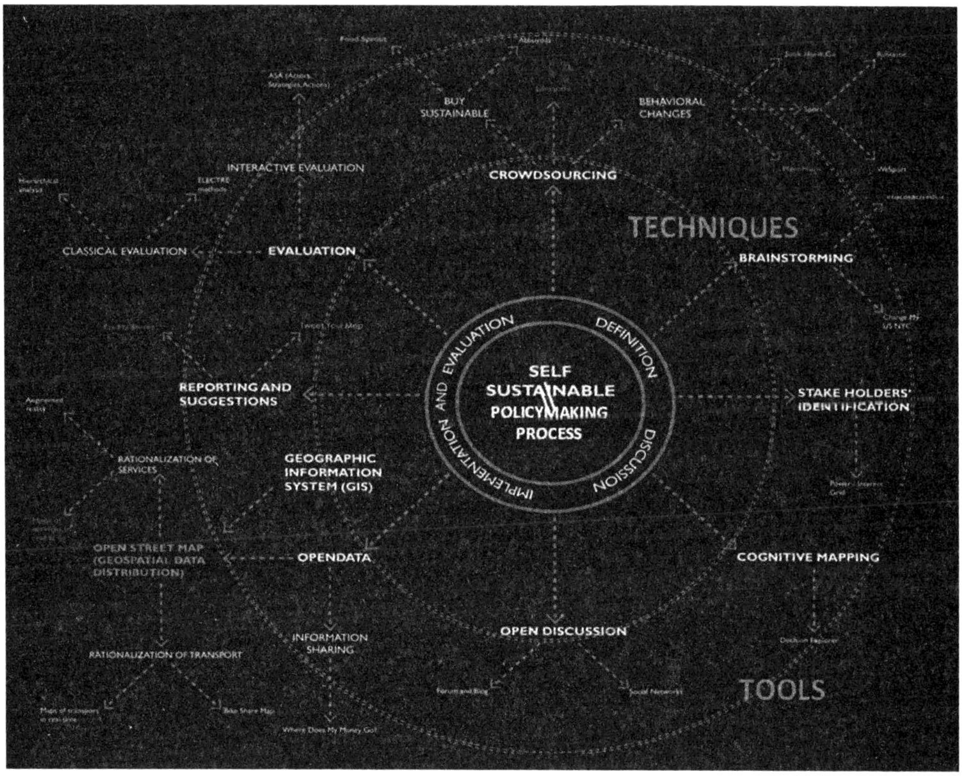

Figure 1 The policymaking process

The problem definition is the recognition of the decision, the process by which are rec-

5 Even the non-programmed decisions, such as strategic decisions, can contain elements of familiarity in the structure, so as to identify the structured processes to take strategic decisions (Mintzberg et al., 1976).

ognized situations that require a response. It concerns fields of collective discussion, such as the environment, services, etc. As Meltsner argues (1976), the definition of a policy problem requires the identification of the actors, the motivations and the beliefs, the resources and the places. Several techniques support decision makers in this first step, e. g. the technique of crowdsourcing. Crowdsourcing is the "practice of obtaining needed services, ideas, or content by soliciting contributions from a large group of people, and especially from an online community, rather than from traditional employees or suppliers".

It combines the efforts of numerous self-identified volunteers or part-time workers, where each contributor of their own initiative adds a small portion to the greater result. For our purpose, crowdsourcing technique can be supported by the use of some specific tools, such as platforms for collecting ideas and opinions: "Ideastorm", "UserVoice", "Ideascale [6]", or some web applications to "buy sustainable" (e. g. "Food sprout" [7]), to manage "behavioural changes" (e. g. "Mom maps" [8], "Junk. Hunt. Ca" [9], "Runtastic" [10] and "We-Sport"). The second technique to cope with the first

---

6 A crowdsourcing platform to bring out the best ideas and to share, vote, and discuss feedback (http://ideascale.com/). One of the most successful use of this software as a service was done in the aftermath of the administrative election in 2011, with the "Ideario per Cagliari" (http://oratoccaanoi.ideascale.com). It was an experience of civic hacking that attracted participants via Facebook and Twitter. The collection system of ideas is very simple and intuitive: users create a profile and once they are member of a community they can submit ideas for the governance of a city. Most popular ideas are prioritized at the top, based on the number of votes the idea receives.

7 FS (Naber & Chang, 2011) collects, aggregates and analyzes information to make the food chain transparent. The interactive maps showing destinations for consumers that want to find sustainable food, where the food come from, know the food impact on the environment and share their knowledge (farmer markets, restaurants, farms) (http://www.foodsprout.com/).

8 MM (Miller, 2009) is a smartphone app that show to find family fun places and kid friendly locations: kid's parks, kid's playgrounds, kid's restaurants, museums and indoor play areas. In addition to a comprehensive play guide, contact info, reviews by parents and times (http://mommaps.com/).

9 JH (Thumb Genius Software, 2010) is a web app that helps the Ottawa residents to reuse items rather sending them to a landfill. It also allows users to upload data.

10 Runtastic (GmbH, 2009) is a GPS application for the personal fitness. It support to monitor the activity (tracking distance, speed, pace, time, heart rate, calorie consumption and route traveled via maps when running, biking or engaging in other exercise activities. It also puts into contact people who practice the same activity or the same routes (https://www.runtastic.com/).

step of problem definition is brainstorming[11]. It helps to bring out ideas aimed at problem solving through users' creativity using whatever means possible (Van Grundy, 1985; Wyatt, 1999). Some customized tools can be useful to manage this step, for example, "Change by Us"[12], a website showing a public noticeboard in the city of New York to allow citizens to share ideas, create projects, discover resources, and make the administration to make the city more liveable. Each user can imagine to participate in a huge on-line "Metaplan", and there is an important section named "who is listening", managed by experts and techno-structures from the administration side. A similar Italian experiment was "E tu cosa ci vedi"[13], developed by the Municipality of Venice in order to produce shared maps (by means of public data, ideas, projects, interviews, etc.) to regenerate an area of the city, currently unused (San Giobbe). Third technique to be considered in the first step is stakeholder analysis, to identify potential networks of actors and stakeholders, such as new coalitions, formal or otherwise, that can be formed to assess the attitudes regarding the potential changes due to a plan or a project. In fact, a cornerstone of EU legislation emphasizes the need for organizations to take into account the interests of everyone, even to justify the decisions taken (Orts, 1992). Tools for SA[14] support decision makers to consider a "list" of stakeholders, broken down to an appropriate level, with the aim to establish a framework of shared strategies to be undertaken.

The phase of the problem definition does not relate to a real accompaniment of participatory procedures by the side of institutions, but it refers to the non-evasiveness of the motivations. This has nothing to do with the discussion on possible alternatives, that address the decisions toward the choice among alternative scenarios. To be participatory a planning process must include interaction, mediation and negotiation in each phase of the process and a continuous accompaniment of these procedures by the institutions. For example, in the deliberative approaches assume considerable importance the discussion

11 Brainstorming is a group or individual creativity technique by which efforts are made to find a conclusion for a specific problem by gathering a list of ideas spontaneously contributed by its member(s). The term was popularized by Alex Faickney Osborn in the 1963 book Applied Imagination (http://en.wikipedia.org/wiki/Brainstorming).

12 Change by US NYC is a new website created by Local Projects (2011) and run by the New York City's Office of the Mayor. It is supported, as partners, by: Citizens Committee for New York City, Pratt Center for Community Development and the U.S. Forest Service (http://nyc.changeby.us).

13 www.comune.venezia.it/etucosacivedi.

14 Mapping techniques as a tool for SA includes: power/interest grid; power/interest star diagram; actor influence network maps and stakeholder rolethink (Eden et al., 1998); three-dimensional grouping of power, interest and attitude (Murray-Webster and Simon 2005) and The Stakeholder Circle (Bourne 2007).

and the debate among participants within the path that will lead to a decision. More correctly, public discussion, rational arguments and constructive debates represent the backbone of every process configured as a deliberative[15]. In this way, the general goal is the integration of technical, administrative and political capacities of the institutions with capacities, knowledge, practices and experience rooted in the context, with the aim to trace a map of the process together (Khakee, 1999). Various group management techniques for representing knowledge and discussion are used in conjunction with different analytical techniques, within successive steps of the planning processes, so as to better manage wicked and ill-defined problems. In this way the computerized techniques used move further away from acting simply as "decision support" tools and move closer towards becoming more proactive, "decision aiding" instruments (Wyatt et al., 2004). There is a vast literature about techniques supporting decision makers in this step of the process, such as: Nominal group technique (NGT) (Delbecq et al., 1975); Delphi method (Dalkey et al, 1963); SWOT Analysis[16] (Humphrey, 1975); Participative Design (Emery, 1993); Deconstruction techniques; Action methods (Fox, 1987; Sternberg et al., 1989); critical systems thinking (CST) (Ulrich, 1983); Strategic Assumption Surfacing and Testing (SAST) (Mason et al., 1981), etc. We consider the testing of the cognitive mapping technique[17] through the software tool "Decision Explorer[18]", that allows to prompt the development of a model by highlighting values, beliefs, and assumptions an individual has about a particular issue (E-

---

15 At the national level of governance, several countries, including Australia, Germany, Great Britain, Ireland, Norway, the Netherlands and the U. S. started innovative experiences inspired by the deliberative approach. In addition to well-known citizen's juries, characterized by the establishment of dialogic processes that involve citizens, experts and facilitators to produce informed verdicts regarding specific controversial issues of public interest, we can mention: consensus conferences, introduced in Denmark in the 80s by the Danish Board of Technology; Conférence de citoyens (http://www.debatpublic.fr), forms of participation organized in France, which allow groups of people to interact with experts on scientific and technological issues, for which there are significant uncertainties and disagreement; Plannungzelle or planning cell, especially in Germany (Dienel and Renn 1985); participatory budgeting, a voluntary process that involves governments and citizens in choices to assign financial resources for the implementation of services and investments (Allegretti, 2003); deliberative polls (Fishkin, 1991). The International Association for Public Participation (IAP2) takes also into account: 21st Century Town Meetings, Appreciative Inquiries, Conversation Café, Citizen Choicework, Sustained Dialogue, Web Lab's Small Group Dialogue, Wisdom Circle, Café World, etc. (Lewansky, 2007).

16 http://en.wikipedia.org/wiki/SWOT_analysis.

17 Such a technique is based on G. Kelly's "Personal Construct Theory" which maintains that subjects see the world through a clear model formed by "constructs", and the personal construction system constitutes people's "construct space" (Kelly, 1955).

18 Decision Explorer Software (1990), 3.3.0 v. (academic), Banxia Software ltd., Kendal, UK (web site: <http://www.Banxia.com>).

den et al. , 1998), and so it captures not only knowledge about the background context but also an understanding of the relationships that underlie such background knowledge. More specifically, a cognitive map (Figure 2) is made up by nodes representing constructs-options, facts, goals and so forth, which are also linked to each other by arrows, so as to form action-oriented chains of argument. Links are inserted between constructs to indicate that one construct "causes" or "may lead to" another one. Moreover, different styles of font, font size and colour are used to distinguish between and display different kind of constructs, such as goals, strategies/key issues, actions and opportunities.

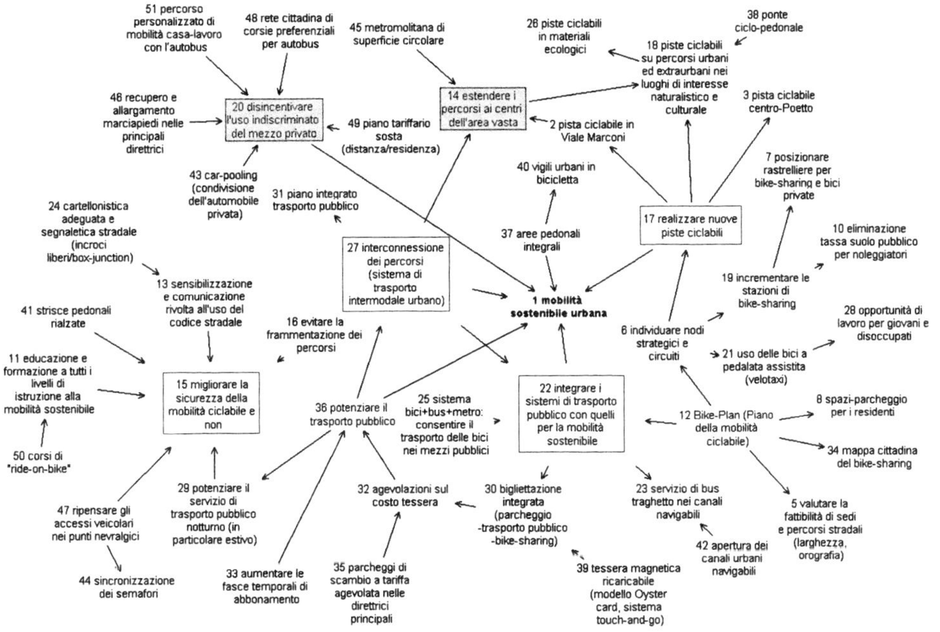

Figure 2 DS tool for representing and analyse knowledge and highlighting key issues: cognitive mapping and Decision Explorer

A number of analyses support the highlighting key issues in the model, by using, for example, cluster analysis, domain analysis and centrality analysis. The latter gives an indication, on the one hand, of the importance of some fundamental themes within the model and, on the other hand, about the necessity of further tests on less represented constructs. Therefore, through investigation about clusters we are able to define which areas of interest and main strategic options are arisen in the interactive space, in addition to the issues with the higher score both in central and in domain analysis.

On another point of view, "open discussion" techniques strengthen networking capacities offered by the internet to support discussion and orient points of view. Just think about the role of social media (Youtube, Facebook, Twitter) in the Northern African revolutions or protest movements in Spain (Indignados), US (Occupy Wall Street) and Greece (Aganaktismenoi) for exchanging information, organising protest events, mobilising participants.

The last step of the process is the implementation and evaluation phase, where we focus on techniques of data collecting, mainly Open data[19], matched with the use of crowdsourcing tools on a geographical basis. At this step, group of ideas are structured and make it usable directly for designing services and facilities. To enable consistent policies choices and actions we need different forms of implementation (and evaluation consequently). We take into account new associations with variable geometry according to the possible scenario put in place. They can involve not only formal institutions, but also public and private entities of various kinds, which can be leaded to identify themselves in a project proposal. This approach can't be embedded in a background of conventional techniques and tools for policymaking. From the implementation side of the process, we focus mainly on several aspect of policymaking and related techniques and tools: representation; sharing information; reporting and suggestion and rationalization of services. New tools should be intuitive, easy to use, and attractive. They must take advantage of the game dimension, as in the Italian platform for urban gaming Critical city. org[20] (Osimo, 2012).

Current model of representation of the city are digital, three-dimensional, multi-resolution and in real time: this is City Model (Borga, 2013). Citysensing[21] is a platform for collecting, processing, and visualizing data about the city and its hidden and secret systems. It hosts collaborative data collection events, collect and process data from sensors, mobile phones, research, and observation, and make it available for anyone to explore. City Model and Sensing grant reciprocal meaning and efficacy, supporting the multi-agent interaction and processes of governance.

---

19 "Open data is the idea that certain data should be freely available to everyone to use and republish as they wish, without restrictions from copyright, patents or other mechanisms of control" (Auer et al., 2007). This concept share the same philosophy with other "Open" movements such as open source, open hardware, open content, and open access.

20 http://criticalcity. org/home/landing

21 http://www. citysensing. org (Normal projects).

GeoMedia[22] is a Geographic Information System (GIS) application for working with and conducting analysis on geographic information as well as producing maps from that analysis. GeoMedia is used for: creating geographic data; managing geospatial databases; joining business data, location intelligence and geographic data together; creating hard and soft-copy maps; conduct analysis in "real-time"; base platform for multiple applications, geographic data validation, publishing geospatial information and analysing mapped information.

Where Does My Money Go? is part of OpenSpending[23], where people can find information about government finance from countries across the world. It enables UK citizens to know how public funds and taxes get spent, starting from the salary. We define here an exemplar e-participation tool that support administrations for conventional aspects, those for which authorities are already required to operate and provide services to citizens. FixMyStreet[24] allows citizens to report the daily problems of their neighbourhood, e. g. water leaking, holes and street drainage, traffic light bent, etc. (Figure 3).

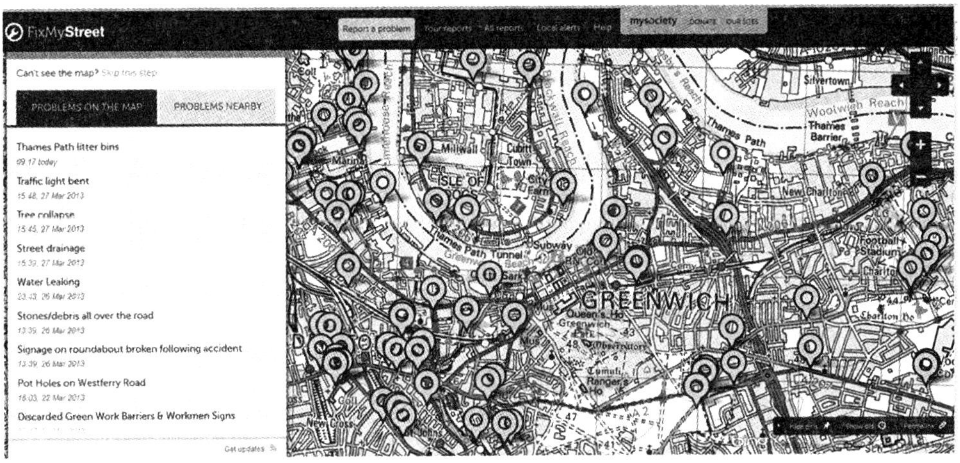

Figure 3 Output from FixMyStreet

This type of instrument has had a large following and countless variations, clones and applications in different fields. OpenStreetMap[25] is a collaborative project, inspired by

---

22 http://en. wikipedia. org/wiki/GeoMedia, developed by Intergraph®.

23 http://wheredoesmymoneygo. org/. This is an Open Knowledge Foundation project, developed with generous support from Channel 4's 4iP.

24 http://www. fixmystreet. com/ is developed by MySociety, a English no-profit organisation

25 Steve Coast, 2004. http://www. openstreetmap. org/

the success of Wikipedia, to create free editable maps of the world, by experiences of the users. Ground surveys are performed by mappers on foot, bicycles, car, boat. They can collect data using GPS devices, aerial photography, and other free sources. These crowdsourced data are then made available under the Open Database License. About the rationalization of public transport, Moovit[26] is a community-driven application for real time public transport information and GPS navigation about plan trips across buses, trams, trains, rapid transit (metro, etc.), along with nearby stops and stations. It integrates and share real time data, crowdsourced by the people, with more static public transport schedules. From the point of view of the accessibility to places and services, to facilitate the opportunities to move and use different forms of urban mobility, BikeShareMap[27] is a GPS web app that shows the network of thousands of bicycles available at self-services docking stations around the city.

There are also e-tools that supports the "walkability" in communities, that is the ability to walk through places, and provide support point-to-point for reaching destination by foot, as RideScout, Walkscore, Walkonomics, and Transit Screen[28], in addition to the well-known Mapquest and Google maps.

According to the idea of Streetscape's president Shaffer[29], administrators "must increase their awareness of mobile solutions that exist (or are in development)" and, in the same time, consider the subset of people who "couldn't care less about technology", by giving information in traditional landmarks "that pedestrian already know to use to get around a city" (Scott, 2014).

As regards the last step of public policy assessment, apart from the traditional families of multi-criteria evaluation methods and techniques such as: multi-objective/goal programming; multi-attribute utility theory methods (Analytical Hierarchy Process-AHP, Multi-Attribute Utility Theory-MAUT); outranking methods (ELECTRE); preference disag-

---

26 http://www.moovitapp.com/ (Tranzmate, 2011).

27 http://bikes.oobrien.com/ (O'Brien, UCL CASA).

28 Walkscore and Walkanomics assess the pedestrian friendliness of an area and then calculate this information into a score. Foursquare is a social media app that helps to find perfect places to go when looking for food, services, entertainment, and other amenities. TransitScreen provides just-in-time information to support walkers to find interconnection with alternative forms of transportation from screens displayed in transit stations and in other locations such as in residential and governmental buildings (Scott, 2014).

29 Streetscape is a Denver-based firm providing pedestrian amenities for public spaces.

gregation methods (UTA); rough set theory methods (Zopounidis et al., 2002), our emphasis is on interactive multi-objective evaluation models for policy selection and prioritization. The software package known as ASA[30] (actors-strategies-actions), for example, allows to compare and prioritize policies depending on three criteria: the interest and the relevance of the options and actions for actors involved, especially synergies among options and actions, and necessary resources. These three criteria are essential for our purpose, as they represent the three fundamental dimensions of public policy decision-making, namely (i) the effectiveness and efficacy of actions (the relevance criterion), (ii) the complexity of and the effects on the political arena (the interest criterion), and finally (iii) the efficiency and resource constraints (the resources criterion).

3 FROM "TOP DOWN" SMART CITIES TO "BOTTOM UP" DIGITAL HAMLETS

To come full circle, an organizational space for interaction is made possible by a web portal with the DSS platform, that provides a valuable interface between citizens, authorities and the private sector and represents solution-oriented innovative approaches in a public and participatory dimension. Furthermore, without forgetting to stress the importance of finding ways to communicate with people who aren't technically adept, this model of development has an eminently innovative and self-organisational dimension as well, that is, with regards to "autopoietic" systems (Maturana et al., 1980), it refers to a system as a network of processes (relations) of internal elaboration of input coming from outside, capable of regenerating and transforming them in stable development energy and social growth. In this way, the computerized techniques and e-tools used move closer towards becoming more proactive, "decision aiding" instruments and the perspectives of each hamlets or districts will depend on the capacities of each one to offer different opportunities based on their own environmental conditions and options.

3.1 A CASE STUDY IN NORTH-WESTERN SARDINIA

In this article, we concentrate on a work-in-progress about an experience of policymaking in a low density residential rural area, named Nurra, a geographical region in the north western Sardinia (Italy). It is the second largest plain of the island, located between the city of Sassari, and the towns of Porto Torres and Alghero. It covers a surface of about 800 km^2 and is bounded by the Sardinian Sea, on the west, and by the Gulf of

---

30 ASA Software (2007), Laboratory of Analysis and Models for planning (LAMP), University of Sassari (web site: http://www.lampnet.org). ASA was tested by one of the authors in several strategic plans. For more technical details see [3]).

Asinara, on the north. Specifically, this area is an explanatory example about changeable relationship between city and countryside. In fact, the region was interested by a huge work of soil drainage during Fascism, that continued also after the World War Ⅱ by the ETFAS Regional agency[31], firstly made by convicts from the Alghero jail, emigrants and, after the war, by refugees. "Despite being next to one of the most populated areas of Sardinia, has one of the lowest population densities in Italy, with 5 inhabitants per km^2"[32], with all related problems, such as: lack of primary and neighbourhood services; lack of people involvement in policymaking, despite a strong social cohesion; lack of an efficient transport system; etc. Close to Alghero, in the countryside marked by canals organized in an orthogonal matrix, there are several rural hamlets: Santa Maria la Palma (1059 inhabitants), Guardia Grande (270), Maristella (424), Tramariglio (8), Sa Segada (395) and Villassunta (20). Fertilia (1703) surely plays a key role in the area and it is also considered a hamlet, despite the fact that it was a Fascist "new town", built in 1936, and it shows many differences from the other hamlets. At the same time, these hamlets are poorly managed and marginal to the historic town of Alghero. By using SWOT analysis we highlight some potential and critical issues to represent the territory, such as: the environmental and social issues, the transport system, the services, the economic structure and the tourism related activities.

As far as the environmental structure is concerned, the environment dominants, as places rich in nature and history structuring and giving strength to the territory, are identified. They are intended as generators of a different kind of settlement space, and, by the way, recognized by the population. In short, the Catch basin of the Calich Lagoon, that consists of natural and sub natural areas on the whole, is afflicted by hydrogeological risk and water pollution, mainly due to high concentrations of phosphorus. Then, the irrigated agricultural areas of the Nurra, that are loosing the settlements values because of unsustainable forms of land-use. Finally, the cliffs of Capo Caccia, part of the Regional Park of Porto Conte and its Marine Protect Area, that show poor interactions between communities and the institution.

Yet, to understand the social structure is important to take into account the socio-demographic analysis. Note that resident population of the area has grown from about 42,000 people in the 1998 to almost 44,000 in 2007. Definitely, these are not large numbers,

---

31 ETFAS, Ente Trasformazione Fondiaria Agricola Sarda.

32 http://en.wikipedia.org/wiki/Nurra.

but, in this region, it is a substantial growth. Special interest have the data on the population of the town of Alghero. In fact, until 2004 the birth rates could still oppose the death rates, even though it was already at lower levels than the national average. Birth rates are also relatively low, tending toward negative natural balances gradually larger. By contrast, as regards the migration rates, it is hard to predict them with a reasonable degree of safety and tolerance, because they have a greater rigidity and can be more easily influenced by external conditions (first of all the economic crisis). In the near future, the migratory component could ensure the population stability and counteract the progressively ageing process and the consequent decrease. In this way, compared to a positive trend of attracting new residents (mainly older people from abroad, in search of a "buen retiro"), we highlight a significant loss of young couples with children, moving in the neighbouring municipalities (e. g. the town of Olmedo), with a more affordable housing market and services. In this context, the rejuvenation of the population is only related to the possibility of constructing policies to attract and maintain the younger population. Hence, the areas relating to neighbourhoods like the historic centre and the more rural areas, are those with the highest rate of older people. The younger population, on the contrary, lives in the countryside and in hamlets, and as was to be expected, in more recent residential neighbourhoods.

As far as the public transport system is concerned, the bus routes are dispersive, unclear and badly managed (lack of information and communication). Although there are a good number of rides, buses travel often empty and do not provide the basic services, needed by the community, so people prefers to move by car. This poor organization of transport public is associated with a total absence of bike routes (except Alghero-Fertilia) and alternative transport systems. Note that the main county roads link Alghero and individual hamlets (mainly the greater ones), leaving the smaller almost completely isolated, like Guardia Grande. Any direct link is provided between the hamlets, just a dense capillary network of connections among hamlets that refers to the mesh of the land reclamation. A more complete transport analysis concerns timing and distance from one hamlet to another.

Nevertheless, the lack of primary and district services in hamlets is crucial (Figure 4): these low density residential areas are detached from the town of Alghero and the few services have unpractical locations, any link with public spaces and poorly managed. For example, the bank and the post office are available only in Fertilia and S. Maria la Palma and the only service offered everywhere is about religious purpose (one church is present in each hamlet).

This is the element that characterizes each hamlet by an individual sense of identity. Fertilia presents some features more similar to a town than a hamlet, with a direct connection to Alghero, and hosts a broad range of services that in most of the hamlets are not present (nursery, primary and secondary schools, police station and emergency medical service, soccer field). Despite a bit more relevant population (and tourist arrivals in the summer season), primary services are not enough and public spaces are limited. S. Maria la Palma hosts different levels schools, public areas and equipped areas for sport, public services (included the library) and commercial activities. It is the largest hamlet, which plays the pin role in the organizational structure of the territory, despite the inefficiency of the links. The other smaller hamlets hosts mainly basic services, e. g. general store and bars, with playgrounds and soccer fields.

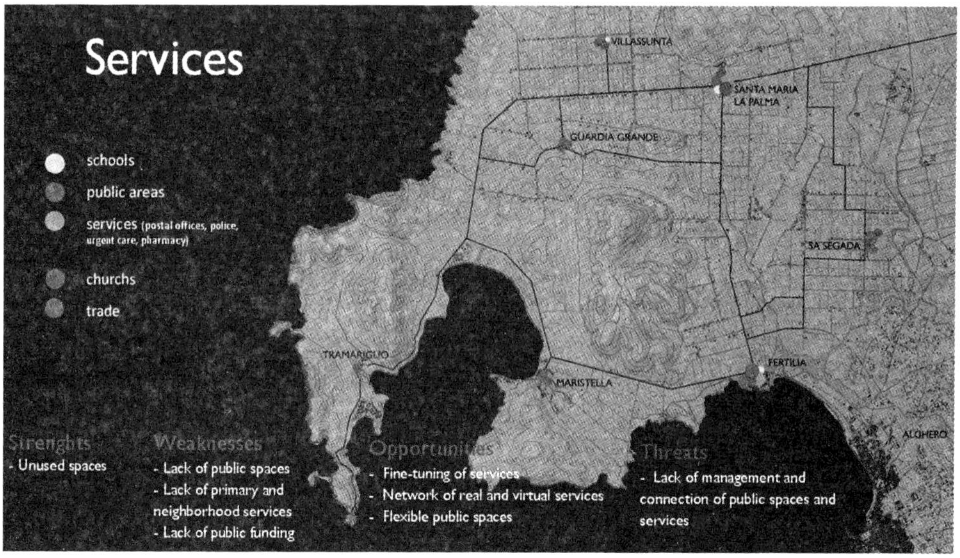

Figure 4 Urban services in the hamlets

As far as the economic structure is concerned, the municipality of Alghero has a level of commercial and industrial activities much greater than the rest of the territory, with a successful agricultural industry; a number of small farmers and a significant tourism industry[33]. The industrial activities, related to agricultural production, refers to the three most important alimentary companies: S. Giuliano olive oil mill, the wineries of S. Ma-

33 In 2010, there were 870,000 international tourist arrivals, within a 3.6 millions as floating population (VI Rapporto sul Turismo ad Alghero ed in Provincia di Sassari, Facoltà di Architettura di Alghero).

ria la Palma and Sella & Mosca, with the related vineyards[34]. To complete, a traditional family-type production, made by small local producers, spreads at the local greengrocers, food markets ("zero food mile") or purchase directly by the customers "on the road". In 2010, the tourism accommodations present in the municipality were: 175 b&b (12 in hamlets, 26 in the area), 32 agritourism enterprises in the hamlets; 3 campings (3 in the area), 35 hotels (6 in hamlets and 4 in the area), 1 hostel (1 in hamlet); 3 tourist village (2 in hamlets and 1 in the area) and 4 holiday houses (4 in the area)[35]. Any specific activity is found for sport, leisure or moving.

3.2 TOWARDS "DIGITAL" HAMLETS

A web portal with the DSS platform manage a mix between interactive tools and real services, working according to an alternative work schedules among hamlets and districts. A DSS increases the effectiveness of the policymaking, as it provides support to the decision-makers coping with problems that cannot be managed by ordinary planning. In our case, the main task of the DSS is to extract useful information for decision-making from a significant amount of data in a short time. Thus, it makes available different techniques and tools, adaptable for specific purposes to the needs of the different users (both residents and visitors) in a "bottom-up" self-sustainable and participatory management system.

First step is the survey of the public spaces in the hamlets with flexible and multi-purpose features along with the survey of the services. The spaces can be chosen because of their strategic location and size. Usually it is better to redevelop unused spaces in a central location, surrounded by green spaces or close to significant element for the hamlet. Thus, each hamlet will have spaces (both indoors and open air) that will represent the node in the network and host services and facilities. The primary services are picked out where weak or even absent: the pharmacy, the register office (with a hypothetical election service for referendum), medical services (immunizations, general practitioner, counselling service), heavy waste disposal, information campaigns (political, social and solidarity) and public information more clear and transparent on water and air quality. A real action plan will allow to identify the alternative services located in various hamlets depending on the day of the week. By contrast, as regards the community services, we choose activities covering different age groups, some will be stable (but not

34 S. Maria la Palma wine cooperative: 326 members; 700 ha; 65.000 q/y; Sella & Mosca wine Spa: Area 550 ha; 55.000 q/y; S. Giuliano olive oil cooperative: 250 ha ; 4500 q/y.

35 Camera di Commercio di Sassari, Comune di Alghero.

open everyday), while others will be itinerant in the hamlets in alternative days: e. g. creative workshops (e. g. practice rooms for music, gaming, writing and ceramics), aggregation centres for the old persons, post-school activities and sports. By contrast, as regards the itinerant activities, we take into consideration the mobile library, the "PlayBus[36]" and "CineBus" so as to provide cultural services in the rural areas to engage young people in different activities, by stimulating creativity and sense of aggregation and to reach people who do not usually have the opportunity to attend libraries. The entire network of libraries (fixed and mobile) will be connected into the library system of Alghero, where people can reserve and get the required books through the "LibraryBus". In the same way, "PlayBus" is a vehicle that weekly transports an entire collection of games in different places of rural or semi-rural areas, such as squares, courtyards, gardens and streets, where groups of volunteers carry out different kinds of animation programs of activities surrounding entertainment and education for children. Hence, the "CineBus[37]" is an example of itinerant cinema for about 30 people that daily move around the hamlets and an itinerant farmer's market, that is present only in town. The design of some spaces by micro-interventions (public lighting, green spaces, pedestrian amenities), where these itinerant activities take place, can make them recognizable and usable even without organized activities. Finally, a bike sharing service (free for residents and low-cost for visitors) and bike paths, linking the hamlets to the places of interest are organized. The web portal and the DSS platform support the management system and the promotion of the network of services as well. In this way, users can book, inquire about events, places and services in advance and perform different activities by linking with the platform. It also will suggest to the users the most useful and available e-tools to match with a specific service (Figure 5, and Figure 6).

To cope with and bridge the technological and cognitive divides, a crucial role is play at the government level. In fact, even if programs does not explicitly fund broadband infrastructure, subsidies are used to upgrade existing telephone networks, so that they are capable of delivering high-speed services, to buy laptops to children residing in poor and

36 http://www.workingonwheels.org. See also "Bookmobile"; "Biobus", a mobile laboratory to learn biology basics (Georgia State University, 1999); "TumbleBus", a school bus converted into a mobile gymnasium (Tumblebus Incorporated, US, 2008).

37 In addition, CineBus uses cinema as an efficient tool to educate and raise awareness among up to 135 rural communities in Senegal. (see also http://plancanada.ca/cinebus-in-senegal-africa-education-to-make-a-difference).

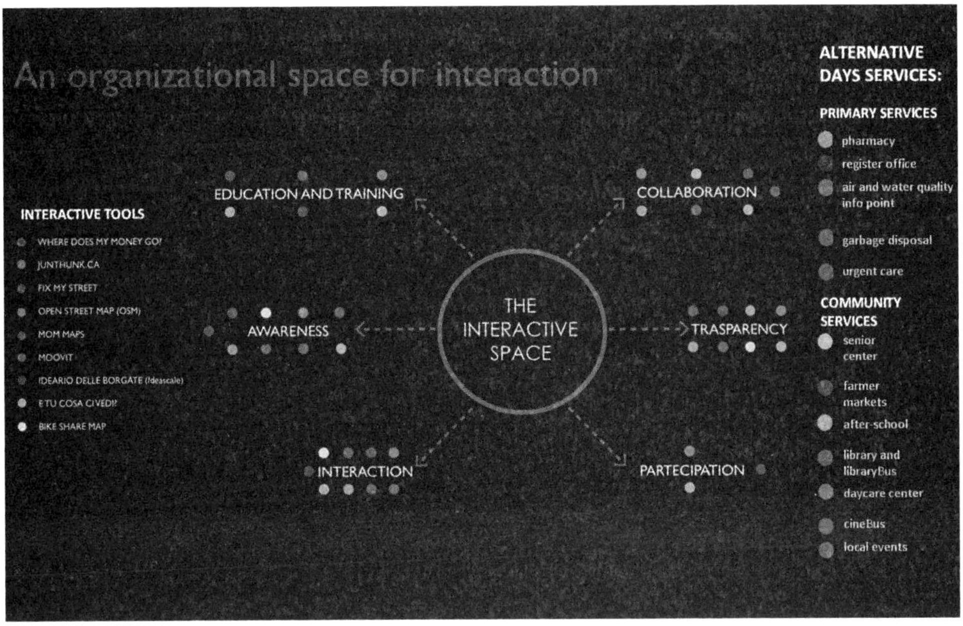

Figure 5 The scheme of DSS platform to organize the space of interaction

isolated regions[38], to incorporate the community members into the process of creating the services. All these actions have to be supported by learning path to gain the basic internet-oriented skills. Thus, for the elderly especially, easy access modes and cultural preparations are very important[39], supported by informal learning and peer group support.

In closing, ICT oriented EU programming must be taken in account to fund the research project. The EAFRD 2014-2020[40] (European Agricultural Fund for Rural Development), in line with the Europe 2020 strategy, agree about some broad objectives of rural development support for 2014-2020, through six EU-wide priorities, including fostering knowledge transfer and innovation in agriculture, forestry and rural areas. Supported actions are related to a package of measures, and Member States may include within

38 OLPC, One Laptop Per Child, 2009.

39 The Berlin "Silver Media" computer senior school "synchronise content and methods according to the need of the target group", offering a general introduction to computers or practical courses like how to store or modify photos, make inventories of collections or use financial software in a mix of education and information, virtual and real communication, entertainment and fun (Schultze, 2003).

40 http://europa.eu/legislation_summaries/agriculture/general_framework/l60032_en.htm. Implementing the national strategic plans is carried out through rural development programmes containing a package of measures grouped around 4 axes. In this case, we consider axis 4, LEADER, for implementation of local development strategies through public-private partnerships called "local action groups".

their rural development programmes the matic sub-programmes, contributing to the Union priorities for rural development, aimed to address specific needs identified, in particular in relation to: on line health care services; small and medium enterprises support; local public services; ICT centres for young people; internet points/public communication; broadband infrastructures.

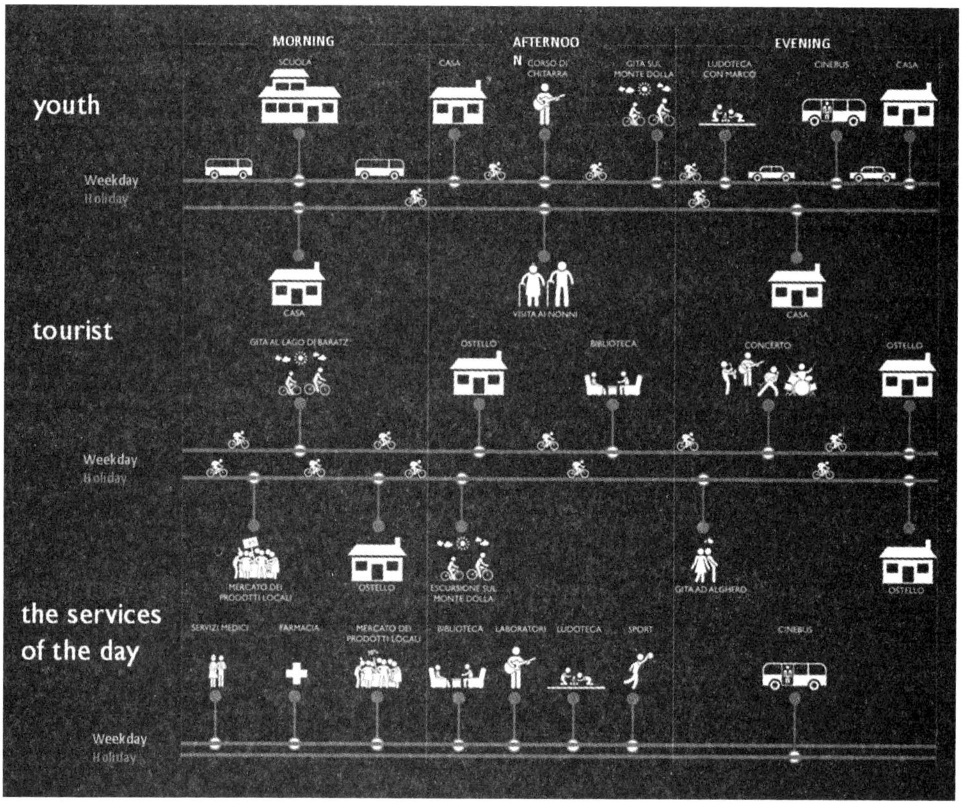

Figure 6 A possible daytime in S. Maria La Palma hamlet

5 CONCLUSION

We conclude that, far from the technology-based smart city paradigm, this bottom-up approach to innovative e-participation practices is extremely useful for three important things. First, it provides an administration with a simple, immediate and free use of a pool of ideas, suggestions and proposals previously unimaginable. Second, it can help local administrations to better and cheaply organize a number of basic services and facilities, to make feel the urban presence especially in rural areas. Specifically, a web portal with the DSS platform manage a mix between interactive tools and real services, working according to an alternative work schedules among hamlets and districts. Third, it fosters citizens and visitors, planning communities and decision making practitioners for the successful implementation of communities development projects.

Against this framework it may be seen, on the one hand, the interaction between analysis and judgment, between processes of social interaction and technology, on the other hand, the integration between technical, administrative and political skills of institutions and techno-structures with knowledge, practices, experiences and needs of local context.

REFERENCES

[1]AUER S R,BIZER C,KOBILAROV G,et al. Dbpedia: a nucleus for a web of open data[J]. The Semantic Web. Lecture Notes in Computer Science,2007(4825): 722.

[2]BLECIC I, CECCHINI A, TALU V. The capability approach in urban quality of life and urban policies: towards a conceptual framework-city project and public spaces, urban and landscape perspectives[J]. Springer,2013 (14): 269-288.

[3]BORGA G. City sensing: approcci, metodi e tecnologie innovative per la Città intelligente[M]. Milano: Franco Angeli, 2013.

[4]COTTICA A. Wikicrazia:governare ai tempi della rete[M]. Palermo :Navarra Editore, 2010.

[5]DICKEY J W. CyberQuest: Problem solving and innovation support system, conceptual, background and experiences[M]. Blacksburg:Ablex, 1993.

[6]EDEN C, ACKERMANN F. Making strategy: the journey of strategic management[M]. London : Sage Publications, 1998.

[7]ENGELBART D C, ENGLISH W K. A research center for augmenting human intellect[C]. Washington DC :The Thompson Book Company, 1968: Vol. 33, Part 1.

[8]KELLY G. The psychology of personal constructs[M]. New York: Norton, 1955.

[9]KHAKEE A. Scenari partecipativi per lo sviluppo sostenibile: temi metodologici[M]. Urbanistica 112, 1999.

[10]LÉVY P L. Intelligence collective: pour une anthropologie du cyberspace[M]. Paris: La Découverte, 1994.

[11]MELTSNER A J. Policy analysis in the Bureaucracy[M]. Berkeley: UC Press, 1976.

[12]MINTZBERG H, RAISINGHANI D, THEORET A. The structure of unstructured decision processes[J]. Administrative Science Quaterley:1976,June:246-275.

[13]OBAMA B. Memorandum for the heads of executive departments and agencies on transparency and open government [J/OL]. 2009,(2011-2-16):http://www. whitehouse. gov/the_press_office/TransparencyandOpenGovernment.

[14]ORTS E W. Beyond shareholders: interpreting corporate constituency statutes[M]. The George Washington: Law Review,1992.

[15]PLAISANT A, VERONA M M. Ideas for a better place: e-participation tools supporting decision making process at the local level[C]. 7th National Conference on Informatics and Planning - Input 2012 proceedings,Cagliari,2012.

[16] SEN A K. Capability and well-being[C]// SEN A K, NUSSBAUM M. The quality of life. Oxford: Clarendon, 1993.

[17] SHIRKY C. Here comes everybody: the power of organizing without organizations[M]. New York: Penguin Press, 2008.

[18] SHIRKY C. Cognitive surplus: creativity and generosity in a connected age[M]. New York :Penguin Press, 2010.

[19] VAN GRUNDY A B. Techiniques of structured problem solving[M]. 2nd. New York: Van Nostrand Reinhold, 1985.

[20] WYATT R G. Computer-aided policymaking[M]. London: E&FN Spon,1999.

[21] WYATT R G, PLAISANT A, SMITH J. Using decision-aiding software and participatory workshops, for better strategic management of a public authority[C]. Grenoble: Congress AESOP Proceedings, July 2004.

[22] ZOPOUNIDIS C, DOUMPOS M. Multi-criteria decision aid in financial decision making: methodologies and literature review[J]. Journal of Multi-Criteria Decision Analysis,2002: 167-186.

Part 3 Managing the Relationship between City and Countryside

Non-Urbanized Areas, Urban Sprawl and Landscape Protection Planning: Researches and Planning Examples from South Italy

Daniele La Rosa
Department of Civil Engineer and Architecture-University of Catania-Italy
dlarosa@ darc. unict. it

ABSTRACT: Non-Urbanized Areas (NUAs) are outdoor places in urban and peri-urban contexts with significant amounts of vegetation, often representing the last remnants of nature in metropolitan contexts. As part of the agricultural and green infrastructure, NUAs provide a different range of Ecosystem Services, such as purification of air and water, mitigation of floods and droughts, regeneration of soil fertility, moderation of temperature extremes, and enhancing of landscape quality. Like other natural ecosystems, NUAs today are endangered by urbanization processes, which are the main causes of their fragmentation and loss of evapotranspiring features. For these reasons, the protection of these areas is a fundamental issue for urban and landscape planning, and it requires appropriate methods and strategies for their management. The protection of NUAs is also needed in landscape protection planning. Particularly, hazard categories such as landscape transformations by new planned developments, intensification of urban sprawl, loss of agriculture land and soil erosion need precise assessment tools, as they can generate a risk of loss of landscape value. The evaluation of these phenomena can provide fundamental information for the definition of Landscape Protection Plans, in which the selection of norms for protection or rehabilitation is strictly related to hazards and risks that are found.

Key Words: Non-Urbanized Areas, planning/designing the landscape, Ecosystem Services, soil consumption, XXI century urbanization processes

1 THE ROLE OF NON-URBANIZED AREAS IN CONTEMPORARY URBAN CONTEXTS

1.1 A FIRST DEFINITION OF NON-URBANIZED AREAS

With reference to metropolitan contexts, Non-Urbanized Areas (NUAs) are part of the urban and (semi) natural areas that produce ecosystem services (La Rosa and Privitera, 2013). They are outdoor places with significant amounts of vegetation, often representing the last remnants of nature (Jim and Chen, 2003). They preserve biodiversity in urban areas (Kowarik, 2011), sequester CO_2 (McHale et al., 2007; Nowak and Crane, 2002), produce O_2 (Jo, 2002), reduce air pollution (Yang et al., 2005) and noise (Fang and Ling, 2003), regulate microclimates, reduce the heat island effect (Shin and Lee, 2005), affect house prices (Kong and Nakagoshi, 2006), have recreational value (Tarrant and Cordell, 2002), and play a fundamental role in health, well-being, and social safety (Groenewegen et al., 2006).

In literature, NUAs have been classified in different ways, resulting in a varied range of definitions (European Commission, 2004). Peri-urban (natural) spaces, spaces free from urbanization, and urban parks are therefore different meanings of areas that in urban contexts are still not urbanized or that maintain their natural or seminatural status. NUAs can include different urban ecosystems, depending on geographical regions. The term "urban ecosystems" can be used to identifies all natural green and blue areas in a city (Bolund and Hunhammar, 1999).

For urban and landscape planning purposes, classifications and typologies of NUAs must be site specific as well as orientated to the aims of the classification itself. However, some common typologies can be outlined, as general families of NUAs that can be found in urban contexts, where a varied range of mixed land uses can be observed (La Rosa, 2012). Farmlands, in use or abandoned, urban and peri-urban forest, shrublands, grasslands, streams and lakes can be part of the heterogeneous fabric of NUAs (Figure 1).

1.2 AN OVERVIEW OF ECOSYSTEM SERVICES PROVIDED BY NON-URBANIZED AREAS

As previously introduced, NUAs provide different types of Ecosystem Services (ES). Ecosystem functions are the physical, chemical, and biological processes occurring in ecosystems that are necessary for its self-maintenance and are the result of interactions between the biotic and abiotic components of an ecosystem (De Groot et al., 2002). According to Daily (1997), these functions are called ecosystem services and are defined as those conditions and processes through which natural ecosystems, and the species that inhabit them, sustain and fulfill human life. From this perspective, ES contribute to human well-being, since they are end products of various ecosystem functions such as climate amelioration and recreation and because they are enjoyed, consumed or used by humans. Ecosystem goods can be defined as tangible products derived by ES, such as wood, fuel, or food as results of ecosystem processes (De Groot et al., 2002).

For urban areas, Bolund and Hunhammar (1999) discussed some of the generated services from the set listed by Costanza et al. (1997), such as air filtering (gas regulation), microclimate regulation, noise reduction (disturbance regulation), rainwater drainage (water regulation), sewage treatment (waste treatment), and recreational/cultural values. Other services, such as food production and erosion control, usually have a lesser importance within cities' limits but may become relevant in more extended contexts such as metropolitan areas.

Air cleaning and pollution reduction

Urban vegetation plays an important role in the environment of cities, reducing atmospheric levels of greenhouse gases and PM10 (McHale et al., 2007). A work by Nowak et al. (2006) about cities in the United States showed that pollution removal varied among cities, with total annual air pollution removal by urban trees estimated at 711,000 metric tons (with a $3.8 billion of value).

Vegetation generally has a greater capacity to filter the air than water or other open spaces and urban forests can directly affect air quality in two different ways: (1) by increasing dry depositions and thus removing air pollution from the atmosphere, and (2) by increasing biogenic volatile organic compound (BVOC) emissions that can act as precursors of secondary air pollutants, though both direct and indirect effects can occur (Escobedo et al., 2011).

Climate regulation

Increased air temperatures can be expected to be particularly problematic in urban areas, where they already tend to be a few degrees warmer than the surrounding countryside. This difference in temperature between urban and rural areas has been called the "urban heat island effect", and it depends on the thermal conductivity and specific heat capacities of materials used in urban areas, the surface albedo, the geometry of urban streets (urban canyons), and the input of anthropogenic heat in urban environments.

Urban greening could affect temperatures through different processes (Bowler et al., 2010). A key process is evapotranspiration, which describes the loss of water from a plant as a vapor into the atmosphere. In addition to cooling by evapotranspiration, shading from trees intercepts solar radiation and prevents the warming of the land surface and air (Oke, 1989). Finally, vegetation may affect air movements and heat exchange. A recent review by Bowler et al. (2010) of different studies about urban greening showed that, on average, an urban park would have around 1°C less than a non-green site.

Control of water runoff

The influence of land use on surface runoff generation is a very complex issue, as land use and land cover have an effect on interception, surface retention, evapotranspiration, and resistance to overland flow. The major effects of urban land uses on surface runoff are caused by the replacement of green features by more impermeable features like buildings, paved surfaces, and roads. These changes dramatically alter the precipitation runoff after rain.

A higher proportion of rainfall becomes surface-water runoff, which results in increased peak flood discharges and degraded water quality through the pickup of urban street pollutants (Haughton and Hunter, 1994). Green areas play an essential role in solving this problem in several ways: in vegetated areas, only 5% ~ 15% of the rainwater runs off the ground, with the rest evaporating or infiltrating (Bolund and Hunhammar, 1999). In cities without vegetation, about 60% of the rainwater is instead led off through storm water drains.

Biodiversity

Natural and semi-natural ecosystems provide the living space for all species on Earth and the maintenance of healthy habitats is directly or indirectly a necessary precondition for the provision of all ecosystem goods and services.

Cities as a whole can be easily seen as novel systems contrasting with rural surroundings, but scaling down to the habitat level shows significant differences between habitats within the city. Kowarik (2011) illustrates major types of ecosystems that can be found in urban areas, ranging from pristine remnants to agriculturally systems, to distinctive urban-industrial ecosystems that emerge mostly after habitat destruction. According to the same author, strategies for biodiversity conservation in urban areas should be aimed at preserving (semi) natural remnants and enhancing native species in urban regions with approaches that acknowledge and don't refuse the contribution of novel urban ecosystems and associated species.

Recreation and cultural values

These kind of services are also called information functions (de Groot et al., 2002). The recreational aspects of all urban ecosystems, with possibilities to play and rest, are highly valued ES in cities (Figure 1). Urban ecosystems (particularly urban greenspaces) provide aesthetic and cultural values to the entire city. These types of services represent nonmaterial benefits that can be obtained from ecosystems through spiritual enrichment, cognitive development, reflection, recreation, and aesthetic experiences.

Farmlands Shrublands

Urban grassland Urban greenspaces

Urban forest in city park

Figure 1 Examples of NUAs in urban contexts

Soil retention and formation

Soil is formed through the disintegration and degradation of rocks, and it gradually becomes fertile through the accretion of animal and plant organic matter and the release of minerals. Soil formation is a very slow process (Rasio, 1999), and natural soils are generated at a rate of only a few centimeters per century and have to face continuous erosion processes. Ecosystem services derived from soil formation relate to the maintenance of crop productivity on farmlands and the integrity and functioning of natural ecosystems. The soil retention function mainly depends on structural aspects such as vegetation cover and roots. The services provided by this function are very important in order to maintain agricultural productivity and prevent damage due to soil erosion (both from landslides and dust bowls).

2 PRESSURES ON NUAS IN URBAN CONTEXTS

2.1 SPRAWL PROCESSES AND NUAS IN CONTEMPORARY METROPOLITAN AREAS

For decades, in many European countries, the dynamics of urban and economic growth have been separated from the demographic development (Kasanko et al., 2006). But despite the decreasing of the population, urban expansion due to spatial development pressure has been an impressive driver of very high consumption of land and agricultural resources. This has resulted in an overall decreasing of the provision of ecosystem services. In the period between 1990 and 2000, at least 2.8% of Europe's land experienced a change in use "including significant increase in urban areas" (Commission of the European Communities, 2006).

The European Environment Agency (EEA) has described the process of the urban sprawl "as the physical pattern of low-density expansion of large urban areas, under market conditions, mainly into the surrounding agricultural areas" (EEA, 2006). Sprawl is the leading edge of urban growth, and it is usually related to limited planning control in land allocation. Urban development is usually patchy, scattered, and strung out, with a tendency for discontinuity. It leapfrogs over areas, leaving agricultural enclaves. Sprawling cities are the opposite of compact cities—full of empty spaces that indicate the inefficiencies in development and highlight the consequences of uncontrolled growth."

In the United States, sprawl concept introduces a number of issues that have been deeply debated in the literature, with controversial results: for instance, Burchell et al. (1998) synthesize 40 years of research on urban sprawl's impacts and conclude that the three conditions that define the negative impacts of sprawl—leapfrog development, low density, and unlimited outward expansion—are the same ones that define positive aspects of sprawl as well (Johnson, 2001).

In Italy, since early 70s, the main cities have started to experience a new model of spatial development that has deeply modified and altered the urban landscape. This process has been referred to in different ways, all related to the metropolitanization of the landscape (Camagni et al., 2002) or the widespread city (Indovina, 1990) and it is present in many metropolitan areas. The relationship between the agricultural landscape and the city has produced a particular contemporary peri-urban landscape, where residential low-density settlements are mixed with farmlands that have been partially modified and reduced by urbanizations. A low-density settlement has widely become the main landmark of new metropolitan areas.

Impacts of sprawl on rural and natural areas

The externalities and impacts of sprawl growth patterns on the environment and landscape have been the focus of several studies. The impacts include the following: the loss of fragile environmental lands, increases in air pollution and energy consumption, decreases in the aesthetic appeal of the landscape, the loss or fragmentation of farmland, a reduction in biodiversity, increases in water runoff and risks of flooding, and ecosystem fragmentation (Galster et al., 2001).

Land sustains many ecosystems functions (e.g., production of food, habitat for species, recreation, water retention, and storage) that are directly linked with existing land uses. Impacts on natural areas are also exacerbated by the increased proximity and accessibility of urban activities to these areas, which in the past were farther from "urban influence." This proximity produces stress on ecosystems and species through noise and air pollution. Moreover, the fragmentation caused by transport infrastructures and other urban-related activities creates significant barrier effects that can degrade the ecological functions of natural habitats. From an ecological point of view, fragmentation can heavily modify corridor spaces for species or can isolate populations by reducing habitats to extend below the minimum area required for the survival of these species.

The loss of agricultural land is often directly connected to land consumption due to sprawl processes (Olson and Lyson, 1999). New urbanizations often occur in proximity to already urbanized areas or existing infrastructure because the price of agricultural land is lower if compared to residential zoned land. Agricultural land usually becomes a highly attractive target for investors and urban developers (EEA, 2006). For these reasons, the hazard of loss of agricultural land may be potentially higher in areas close to already urbanized lands or roads. According to the EEA (2006), urban development tends to "consume the best agricultural lands, displacing agricultural activity to both less productive areas (requiring higher inputs of water and fertilizers) and more remote upland locations (with increased risk of soil erosion)."

Social and economic cost of sprawl

From a social point of view, sprawl can generate segregation of residential development according to the higher degree of income that is observable in some sprawled settlements, even if this aspect is less significant today than it was in the past and can be observed with less intensity, at least in some Italian metropolitan areas. The socioeconomic profiles in suburban and peripheral areas are typically middle and upper-income families with children, who require the necessity of high mobility for their daily activities. More and more people are moving away from the center of metropolitan areas, apparently attracted by the quality of life in these rural settings, to live in residential developments built on converted peri-urban farmlands. "The detached terrace-houses and semi-detached houses condense the new type of residential landscape in the metropolitan peripheries of the cities of southern Europe." (Munoz, 2003). From a social perspective, the different issues concerning the contemporary sprawled models and their impacts are very difficult for citizens to communicate about and understand: the model of detached houses is still very idealized and desirable, with a general belief that living far from the city center may be more healthy, pleasant, and "natural" than living within the urban areas.

From an economic perspective, urban sprawl is costly form of urban development than more compact ones for a number of reasons (Burchell et al., 1998; Deal and Schunk, 2004; Zolnik, 2012), such as:

- increased time spent on commuting from home to work over longer distances;

- costs generated by inefficient transportation systems;

- additional costs of building new urban infrastructures and related services;

- reduction of the agricultural value of soils with high proximity to urban areas;

- reduction of incomes from farmers, caused by the reduction of farmland size or their fragmentation;

- greater energy requirements for sprawled settlements as compared to more compact settlements.

Urban sprawl also tends to inhibit the development of public transport based on the development of mass transportation systems: the provision of alternative choices in transportation are essential to ensure the efficient working of contemporary urban environments.

Aspects of economic inefficiency are also associated with the market-oriented urban planning that has frequently generated sprawled settlements and thus has had some responsibility in sprawl processes. This kind of real estate market pushes for urban development of new areas without taking into account the potential reuse of former residential or industrial areas.

2.2 WHAT'S LEFT

What's left today of the peri-urban natural and agricultural areas? A different mix in types and sizes of residual and Non-Urbanized Areas deeply characterize metropolitan landscapes in many European regions, such as farmlands (abandoned or still in use), small orchards, wood and shrubs areas, local parks, regional parks, reserves and natural protected areas, and grasslands. These are the typical features that can be found present in metropolitan contexts (Figure 2).

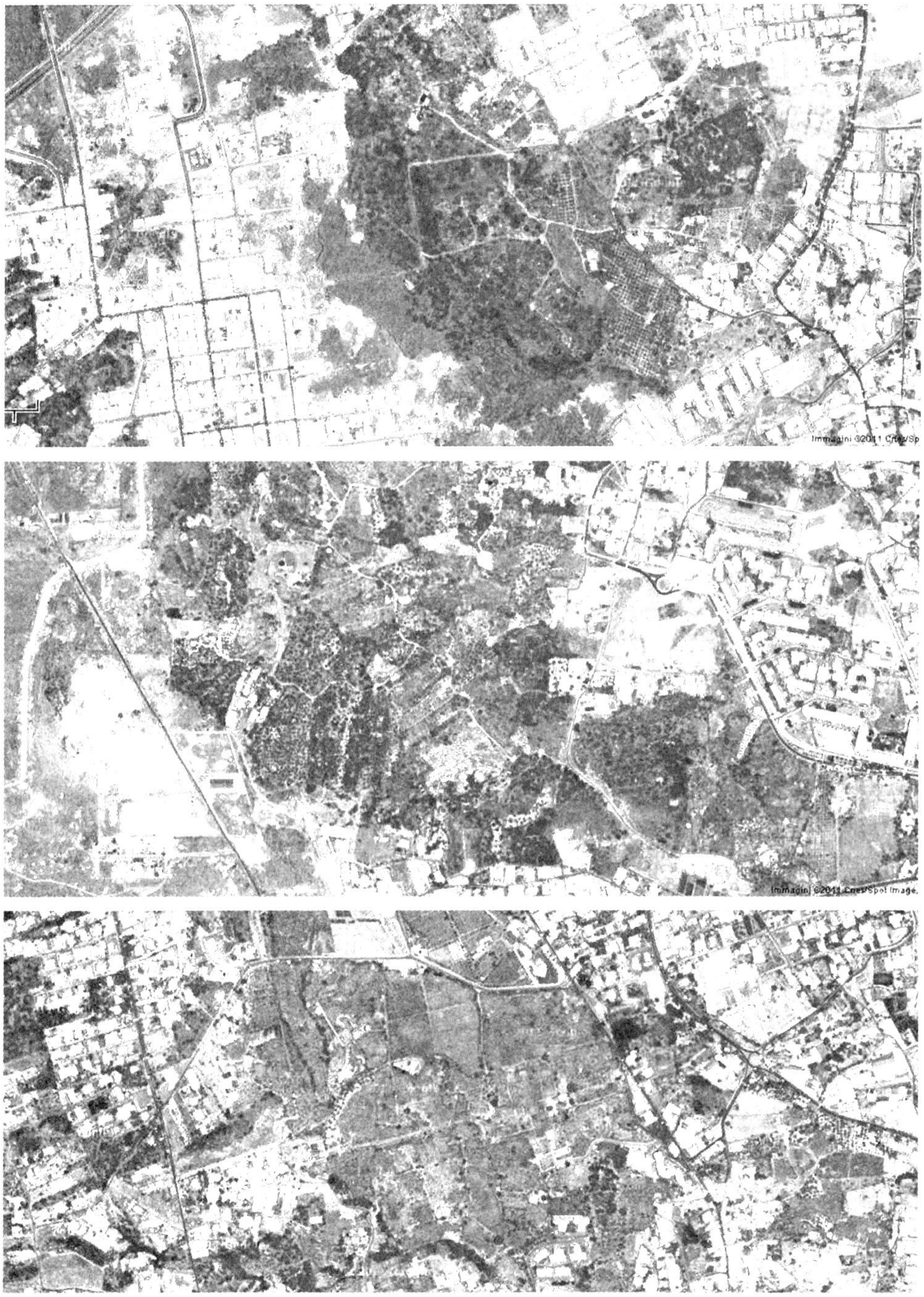

Figure 2 Examples of Non-Urbanized Areas of different type and size in the Catania Metropolitan area, Italy (from Google maps)

Gallent and Shaw (2007) identified a number of land uses in the transition zone from urban to rural of the greenbelts in the United Kingdom: service functions, commercial activities; noisy and unsociable uses pushed away from people; transient uses such as markets; bulk-retail; light manufacturing; warehousing and distribution; some public institutions; degraded farmland; planned recreational areas such as country parks; fragmented residential development (often centered on road junctions); areas of unkempt rough or derelict land awaiting reuse. These pattern of land uses are very similar to the ones that can be found in other European contexts, with various ranges of size and extent.

No consideration about natural resources (e. g. soil, water, species, landscape) has usually been attributed to these areas. Different generations of urban plans have used soil and agricultural lands as unlimited resources, without any constraint except their geographical extension. In the past, urban planning (strongly driven by the real estate market) has not been able to evaluate agricultural areas in a proper way so as to recognize the roles, functions, and services they provide to humans.

This indifference of urban planning toward agricultural peri-urban areas has been one of the reasons at the base of the contemporary un-regulated expansion and sprawling of urbanization process. Since the land had no particular inner value, planners and city decision makers have been used to design new developments without taking into account the peculiarities and features of peri-urban and agricultural landscapes.

Despite the dramatic urbanization processes, NUAs are still present in metropolitan contexts, and they still provide important and numerous ecosystem services and thus need to be accurately analyzed and assessed in order to build new scenarios for their land use. Moreover, understanding the role of contemporary NUAs is important for a correct and up-to-date analysis of the development of metropolitan contexts and the complex processes that have produced them.

3 METHODOLOGICAL TOOLS FOR THE PROTECTION OF NUAS IN URBAN AND LANDSCAPE PLANNING: EXPERIENCES FROM SICILY (ITALY)

As urban areas are expected to keep growing in the future, planners and political decision makers have to carefully consider the role of NUAs that provide ecosystem services. A better knowledge of the different features of NUAs allows us to identify the more suitable land uses to fulfill the aims of conservation and leisure as well as the promotion of new forms of agriculture (La Greca et al., 2011). Land-use planning may allow the protection of green areas that have evapotranspiring and permeable features. This action is directly related to adaptation to climate changes because it can reduce the urban heat island effect (Bowler et al., 2010) and excessive rainwater runoff. NUAs are also fundamental to increasing urban environmental quality by creating more pedestrian-friendly and visually pleasant settlements.

The characterization of NUAs with appropriate analytical tools is therefore a fundamental step in identifying their peculiarities and potentialities and better choosing the most appropriate land uses to maintain their integrity and provided ecosystem services. Different models and methodologies can be applied in order to help the planning process to better address a new scenario of land uses for these areas.

In Europe, agricultural landscapes have a long history that reflects their historical and cultural background. Changes in the spatial patterns of contemporary landscapes are an expression of the dynamic interaction between natural and cultural forces in the environment (Antrop, 2004). Landscapes have been transformed and used according to the needs of communities; this has resulted in a simplification of landscape patterns with impacts on ecological values and biodiversity (Millennium Ecosystem Assessment, 2003). In European agricultural landscapes, over the last fifty years, agricultural practices have intensified, urban and industrial land development have sealed soils, construction of transport networks have fragmented natural environments. All these processes have resulted in deep change of landscapes and land-use, often producing land fragmentation and loss of diversity. Particularly, urbanization processes are the main driver of agricultural landscape transformation in Europe (European Environmental Agency, 2006). In Italy, a recent study (Sansa et al., 2010) assessed the overall rate of urbanization at 244,000 ha per year from 1990 to 2005. Land consumption due to urban sprawl is one of the most widespread hazards for landscapes. Land consumption by urban sprawl and relative loss of agricultural land area among the most endangering hazards that were observed and assessed with the available geographical data in the study area. Landscape planning can play an important role in conserve and/or strengthen such a vast set of services by implementing a set of norms and indications constituting a binding rule for landscape transformations. Norms and indications must be based on sound assessment methodologies, able to characterize existing values and hazards that are present in the landscape.

In this section, two methodologies for the protection of NUAs against the aforementioned pressures are introduced. The first refers to a urban scale, the second to a regional scale. Both were elaborated for real planning experience in two areas of Sicily (Italy).

3.1 THE CHARACTERIZATION OF NUAS FOR URBAN PLANNING OF METROPOLITAN

A better knowledge of the different features of NUAs allows us to identify the more suitable land uses to fulfill the aims of conservation and leisure as well as the promotion of new forms of agriculture (La Greca et al., 2011).

The characterization of NUAs is thus aimed at better choosing the most appropriate new alternative land uses to maintain their integrity and provided ecosystem services. The methodology proposes a five-step process to characterize NUAs and identify possible new land uses to maximize the relative ecosystem services provision. It is tested on three municipalities in Sicily, Italy. This is a settlement system nowadays mainly characterized by a considerable amount of urban sprawl and that experienced a population growth of more than 27% in the last 40 years, following a general trend of loss of the population of the main city (Catania, eastern Sicily) and increase of the surrounding municipalities (La Greca et al., 2011).

The first step in the process is to map the land uses in the entire area, using existing cartography and high resolution orthophotos including categories such as agricultural, road and residential use.

However, traditional land use categories are generally not sufficient to provide indication about land cover features such as presence of trees and relative degree of evapotranspiration (Brown et al., 2000). A patch of a specific land use type can be composed by a complex mix of land covers, so that the identification of different land covers becomes a crucial analytical phase. A land cover analysis of existing land use type based on a geographic sampling combined with a field experience approach was carried out at a very detailed scale, in order to identify those land cover surfaces (i.e. trees, small grass or urban gardens) within the different land use types. This allows the average "evapotranspiration" of each land use to be estimated, a cooling effect caused by evaporation from the soil and the movement of water through plants.

To further describe NUAs, the method also proposes an index of ecological fragmentation, which indicates the size of the NUAs and the number of other NUAs within 500 meters. This index is important, since some land uses, such as playgrounds, might be more suited to small patches of land spread out across the city or town.

A further index of proximity is than introduced, accounting for the number of inhabitants able to have access to the areas. This measure allows to understand how local people might use NUAs. For example, children or older people need to walk small distances to get to playgrounds or allotment gardens. One explanation underlying the "attractiveness" of a land use is the utility associated with the activity to be conducted at the destination. This notion is particularly relevant in urban settlements characterized by high population density, the presence of multi-story apartments and the lack of public and private gardens. Specific land uses may be preferred for these areas, especially those characterized by a high proximity to residential areas. For instance, allotment gardens or playgrounds are typical urban land uses that feature small patches with a good proximity to residential areas. Traveling to activities in these areas usually requires short time spans and distances.

Combining the results coming from the previous analyses, a first option for new land uses of NUAs is proposed in a Land-Use Suitability Matrix (La Rosa and Privitera, 2013). According to the intersection of results from previous analyses, nine new land uses for NUAs have been introduced (Figure 3): Natural parks, Agricultural parks, Community supported agriculture, Allotment gardens, Informal recreation areas, Playgrounds, Friche, Local urban parks, Urban gardens. For example, areas with high evapotranspiration and low fragmentation may be converted to natural parks, but lower evapotranspiration and high fragmentation might be more appropriate for urban agriculture projects.

The proposed new land uses were assessed for compatibility with reference to existing land uses: for example, abandoned farmland would be more easily and productively converted to urban agriculture projects than natural parks. A Compatibility Matrix is used to check the compatibility of current land uses and PLUs. Figure 3 maps the final spatial configuration of new land use of NUAs for the study area.

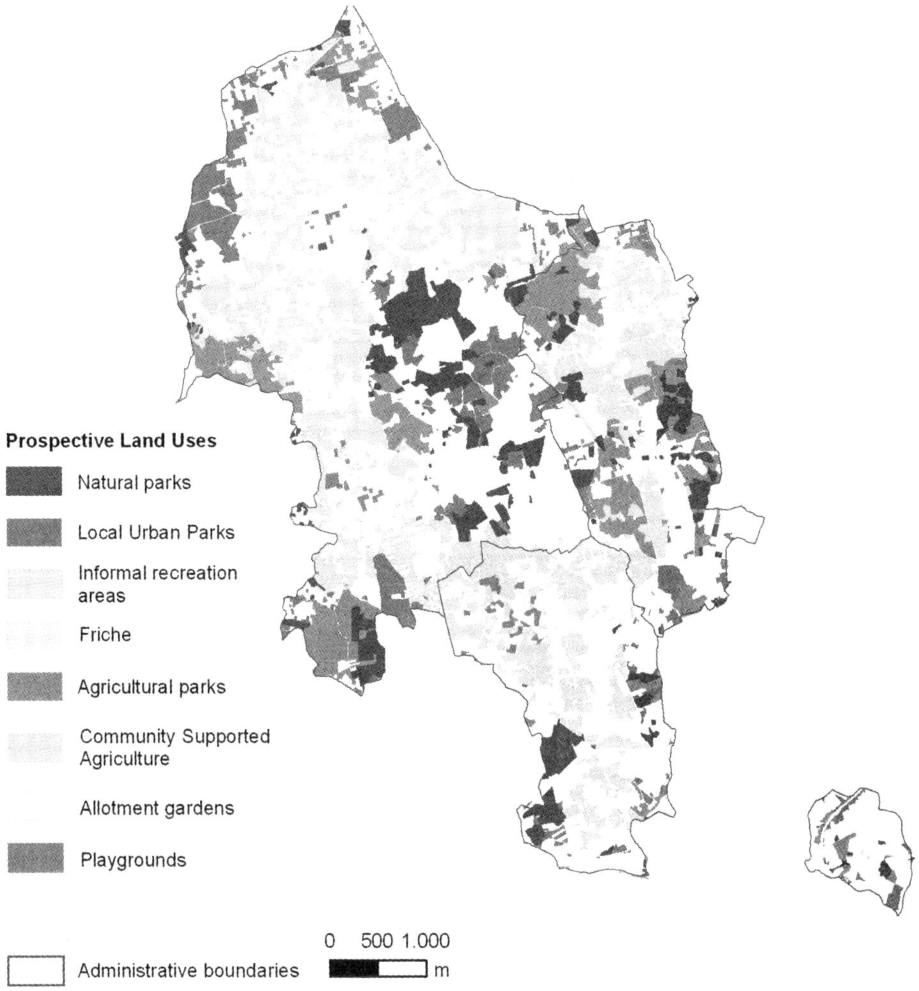

Figure 3 Example of a new scenario of Land Uses for NUAs

3.2 ASSESSMENT OF HAZARDS AND RISKS FOR LANDSCAPE PROTECTION PLANNING

The research here proposes a methodology for the assessment of landscape hazards, values and risks, required by the national legislation of landscape protection planning in the province of Enna, Sicily, where the creation of a Landscape Protection Plan is a compulsory duty of the Regional Department for Cultural and Environmental Heritage and Public Education (La Rosa and Martinico, 2013). The assessment identifies areas subject to different forms and degrees of hazards, landscape value and risks. Results have been used to highlight areas that should be considered more carefully when defining norms and regulations for the landscape. These norms represent a binding reference for all other land-use plans including natural parks, reserves and other municipal master plans.

The selection of assessment variables generally reflects the perceptions and values of the involved stakeholders, and the chosen model is strictly dependent on the systems being assessed and on the aims of the model itself (Gulinck et al., 2001). Consequently, it is not always possible to identify a method that is suitable for all cases, resulting in many assessment frameworks that are generic rather than specific to a particular system or management area.

In the proposed methodology, landscape values were here defined as aesthetic, natural, historical or archaeological qualities of the landscape features. Landscape features were intended as the elements that form the landscape, such as rivers, geological formations, woods, farmlands, archeological areas, etc. The qualities of each feature were evaluated by experts, allowing to highlight the most relevant areas for consideration of appropriate protection norms and restrictions of land uses. For this reason, we focused on integrated expert-based assessment, where experts from different scientific fields provided a scale-based score of the landscape features value (Daikin, 2003; Vouligny et al., 2009). Different value scores of landscape were provided by experts based on their knowledge of the landscape qualities for the features they were asked to assess. Accordingly, numerical scores are intended as a way for measuring the convergence of the experts' evaluations. Results of the assessment (Figure 4) showed that there are not "no values" areas in the landscape and that "low values" as well as "high values" areas must be highlighted for the correct definition of sound planning norms (Gulinck et al., 2001).

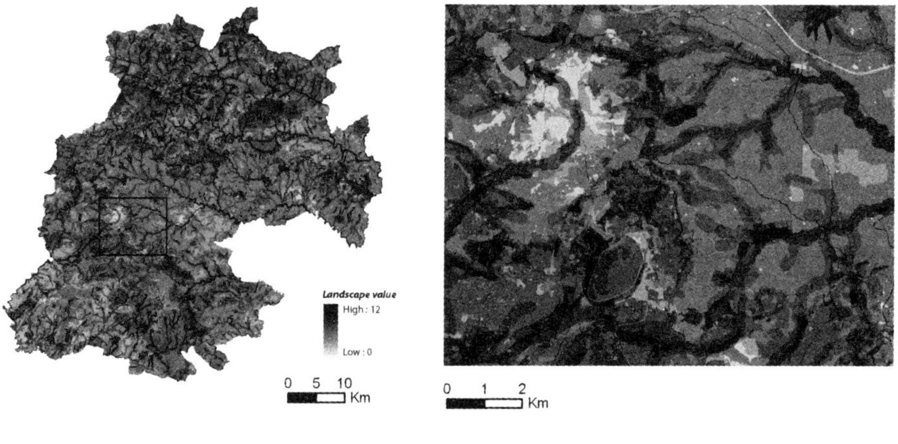

Figure 4 Landscape value mapping for the Province of Enna (Italy)

Although widely used in many fields, such as geology, hydrology or industrial production, Hazard assessments for landscape research appears to be a more uncertain and less explored issue, where approaches have focused on the visual impact of changes in land-use due to human activities (Fernandes, 2000). In this work, hazards were defined as potential threats that are likely to cause damage to the landscape in case they may occur.

Four categories of landscape hazards were identified, considering the ongoing processes of land consumption by urban sprawl and erosion in the study area, the geo-data set availability and the main aims of the Landscape Protection Plan: 1) landscape transformation by new planned developments; 2) intensification of existing sprawl patterns; 3) loss or transformation of agricultural land; 4) transformation of landscape due to erosion. The assessments were performed by GIS functions using both vector geoprocessing and raster analyses.

The final step was the quantification of the risk of loss of landscape value that can occur because of one or more of the hazard categories. It was performed adopting an approach that can be referred to the risk triangle defined by Crichton (1999) that identifies risks as a combination of hazard, vulnerability and exposure. In our approach, the exposure is coincident with the landscape value, while vulnerability is defined as the propensity or inclination of a valuable landscape feature to be damaged by the occurrence of an hazard. An example of the risk of loss of value by loss of agricultural land is shown in Figure 5.

4 DISCUSSIONS AND CONCLUSIONS

A sound characterization of NUAs better addresses urban planning choices

As already discussed, in metropolitan areas NUAs suffer from a wide range of pressures by urbanization processes. The impact and externalities of these processes encompass physical, environmental, socio economical, and political issues (EEA, 2006; Sansa et al., 2010). In some geographical contexts of southern Italy such as Sicily, NUAs have always been targeted by urban plans as generic farmlands or undefined greenspaces without a particular importance or role. This is one of the reasons why agricultural and greenspaces have been continuously eroded by urban sprawl. Weak environmental policies have been one of the results of the lack of attention to NUAs. For this reason, the characterization of NUAs within urban planning is fundamental for the conservation of urban ecosystem services.

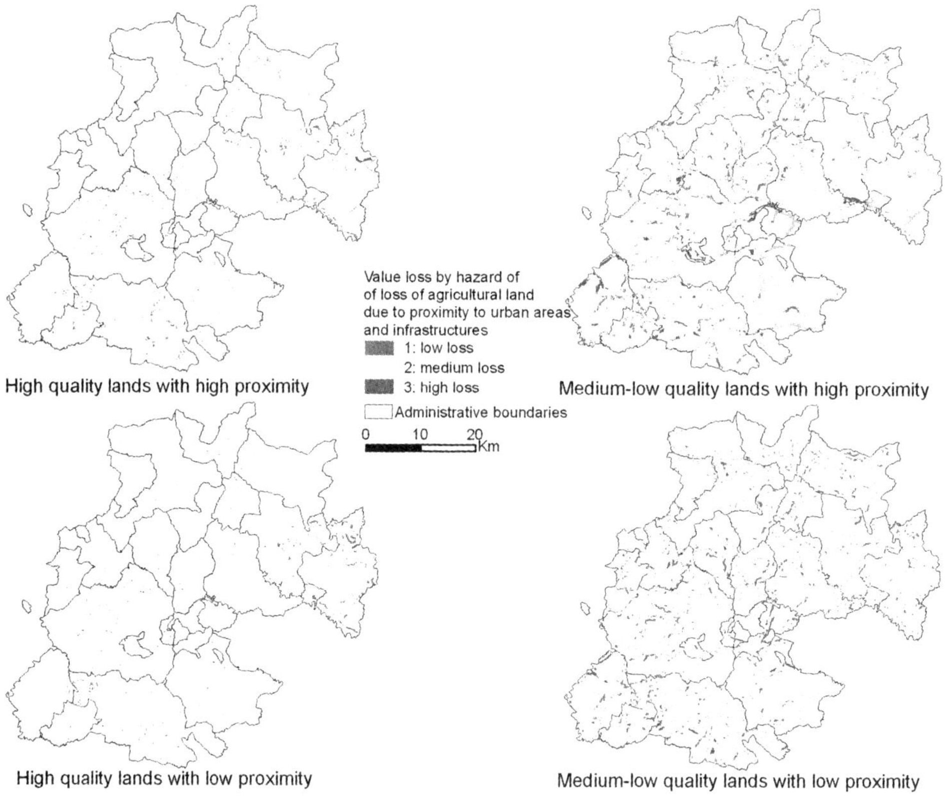

Figure 5 Map of risk of landscape value by hazard of loss of agricultural land

Urban planning strategies for NUAs should be related to the urban surroundings (Hostetler et al., 2011). Therefore, metropolitan areas appear to be the most appropriate scale for rethinking and readdressing land uses of NUAs. Moreover, preserving the components of periurban ecosystems, such as landscape aesthetics and recreational uses, is also important, because they provide more intangible services than other ecosystems (Vejre et al., 2010).

The importance of the method for the characterization of NUAs proposed in the previous section is that it integrates different analytical and evaluating phases to define a new scenario of land uses for NUAs. These analytical phases come from different scientific fields (ecology, remote sensing, GIScience) but can be integrated together in a Planning Support System in order to provide information about particular features of NUAs and to build a new scenario for their conservation. Moreover, the method, as structured, can be expanded and enriched with other analytical phases which can be chosen in order to allow the description and assessment of new features of NUAs (e. g. soil permeability, vegetation richness and rarity).

Moreover, the characterization of NUAs is a base tool for enhancing the current land-use assets, because the proposed scenario improves the overall provision of ecosystem services in the studied area. The proposed new land uses for NUAs might be able to provide a highly differentiated picture of a new spatial configuration of NUAs. Urban plans for NUAs must be developed "within the fuzzy framework of an open space network which is dynamic in aesthetic and ecological status, allowing for a larger mosaic, a patchwork of changing, loose-fit landscapes" (Thompson, 2002). This means that urban plans, at a local or metropolitan scale, have the chance to choose among different possibilities of new land uses and policies. This can be very important, especially for the small local municipalities that have restricted financial resources for projects on green areas.

Integrated assessment of hazards, values and risks allows for defining a graduated set of norms and regulations in landscape planning

The major opportunity for both maintaining ecosystem services and diversity for not directly protected areas such as agricultural landscapes might rely in the promotion of diversity of land uses at the landscape scale. Landscape planning and relative norms have to foster a diversification in land uses and diversity among land users (Swift et al., 2004). Landscape plans should thus include different norms and safeguard rules for the landscape, as well as proposing indications for the rehabilitation and upgrading of urban and agricultural landscapes. The landscape assessment phase can provide fundamental information for the definition of a Landscape Protection Plan, in which the selection of norms for protection or rehabilitation is strictly related to hazards, values and risks that are found.

Current literature highlights the considerable variety of approaches to landscape evaluation (Mander and Uuemaa, 2010). Accordingly, all assessment approaches should be linked to the goals and data availability (Von Wiren-Lehr, 2001). A proper definition of goals, indicators and evaluation methodologies will result in more effective management of landscape norms.

The method discussed in the previous section proposes a detailed landscape assessment over a large area using a limited set of information and integrating different typologies of spatial data. It produces geographically explicit pictures of hazards, value and risk in order to enable the Landscape Protection Plan to propose a varied set of norms and regulations (La Rosa and Martinico, 2013). The method can be also de-composed in submethods (hazards and value assessment) that can be used independently. For instance, the evaluation of the hazard of loss or transformation of agricultural land due to urbanization can be used to assess the impact of different scenarios of urban development on agricultural systems. The structure of the value assessment might allow the inclusion of other typologies of values (i. e. target species assessments) by adding additional data layers.

Even if the assessed hazard categories do not encompass all the different hazards potentially threatening the landscape, they focus on the most relevant on-going processes that are directly menacing many Italian landscapes: land consumption, urban sprawl and land erosion. Results showed that erosion is dominant among the introduced hazard categories, but its index values are also very concentrated on low scores. The hazard of loss of agriculture land is the second most widespread category, but it affects mostly low quality lands. From assessment of the risk of landscape value loss, it can be seen that the most valuable areas of the study area are quite concentrated and are less exposed to risk. On the contrary, more exposed areas are the ones where landscape values have medium and low scores. This result confirms that a key issue for landscape planning is not only the protection outstanding places but also to find appropriate land-use regulations for medium-low quality landscapes, where landscape transformations have been concentrated or are more likely to occur.

It is important to highlight that all assessment methodologies should be intended as tools for orienting landscape planning choices but should not be considered as automatic devices producing prescriptive maps: landscapes are too complex systems to be planned without a wise evaluation by an expert. Nevertheless, all available geo-referenced information represents an auxiliary device, which is helpful for speeding up the process and for managing the overwhelming amount of features that affect the landscape. Particularly, assessment tools based on geographic information system are able provide a systematic and upgradable record of value judgments, major hazards and risks present in a certain area.

REFERENCES

[1]ANTROP M. Landscape change and the urbanization process in Europe[J]. Landscape and Urban Planning,2004,67: 9-26.

[2]BOLUND P, HUNHAMMAR S. Ecosystem services in urban areas[J]. Ecological Economics, 1999,29: 293-300.

[3]BOWLER DE, BUYUNGALI L, KNIGHT T M, et al. Urban greening to cool towns and cities: a systematic review of the empirical evidence[J]. Landscape and Urban Planning, 2010,97: 147-155.

[4]BURCHELL R W, SHAD N A, LISTOKIN D, et al. The costs of sprawl-revisited[R]. Transportation Research Board Report 39. Washington: National Academy Press, 1998.

[5]CAMAGNI R, GIBELLI M C, RIGAMONTI P. Urban mobility and urban form: the social and environmental costs of different patterns of urban expansion[J]. Ecological Economics,2002,40: 199-216.

[6]COSTANZA R, D'ARGE R, DE GROOT R, et al. The value of the world's ecosystem services and natural capital[J]. Nature,1997,387: 252-259.

[7]CRICHTON D. The risk triangle[C]// In Ingleton J(Ed):Natural disaster management[M]. London: Tudor Rose, 1999.

[8]DAILY G, (ed). Nature's services: societal dependence on natural ecosystems[M]. Washington: Island Press, 1997.

[9]DAIKIN S. There's more to landscape than meets the eye: towards inclusive landscape assessment in resource and environmental management[J]. Le Geographe Canadien, 2003,47:185-200.

[10]DEAL B, SCHUNK S. Spatial dynamic modeling and urban land use transformation: a simulation approach to assessing the costs of urban sprawl[J]. Ecological Economics, 2004,51:79-95.

[11]DE CROOT R S, WILSON M A, BOUMANS R M J. A typology for the classification, description and valuation of ecosystem functions, goods and services[J]. Ecological Economics,2002,41: 393-408.

[12]FANG C F, LING D L. Investigation of the noise reduction provided by tree belts[J]. Landscape and Urban Planning, 2003,63: 187-195.

[13]FERNANDES J P. Landscape ecology and conservation management—evaluation of alternatives in a highway EIA process[J]. Environmental Impact Assessment Review, 2000,20: 665-680.

[14]GALLENT N, SHAW D. Spatial planning, area action plans and the rural next urban fringe[J]. Journal of Environmental Planning and Management,2007,50: 617-638.

[15]GALSTER G, HANSON R, RATCLIFFE M R,et al. Wrestling sprawl to the ground: defining and measuring an elusive concept[J]. Housing Policy Debate, 2001,12: 681-717.

[16]GROENEWEGEN P P, BERG A E, VRIES S,et al. Vitamin G: effects of green space on health, well-being, and social safety[J]. Study Protocol, BMC Public Health, 2006,6:149.

[17]GULINCK H, MUGICA M, DE LUCIO J V,et al. A framework for comparative landscape analysis and evaluation based on land cover data, with an application in the Madrid region (Spain) [J]. Landscape and Urban Planning, 2001,55: 257-270.

[18]HAUGHTON G, HUNTER C. Sustainable cities, regional policy and development[M]. London: Jessica Kingsley, 1994.

[19]HOSTETLER M, ALLEN W, MEURK C. Conserving urban biodiversity: creating green infrastructure is only the first step[J]. Landscape and Urban Planning, 2011,100: 369-371.

[20]INDOVINA F. La città diffusa[M]. Venezia: Quaderni DAEST, 1990.

[21]JIM C Y, CHEN S W. Comprehensive green space planning based on landscape ecology principles in compact Nanjing city, China[J]. Landscape and Urban Planning, 2003,65: 95-116.

[22]JO H K. Impacts of urban green space on offsetting carbon emissions for middle Korean[J]. Journal of Environmental Management, 2002,64: 115-126.

[23]JOHNSON M P. Environmental impacts of urban sprawl: a survey of the literature and proposed research agenda[J]. Environment and Planning A,2001,33: 717-735.

[24]KASANKO M, BARREDO J I, LAVALLE C, et al. Are European cities becoming dispersed? A comparative analysis of 15 European urban areas[J]. Landscape and Urban Planning, 2006,77: 111-130.

[25]KONG F, YIN H, NAKAGOSHI N,et al. Urban green space network development for biodiversity conservation: identification based on graph theory and gravity modelling[J]. Landscape and Urban Planning, 2010,95: 16-27.

[26]KOWARIK I. Novel urban ecosystems, bio diversity, and conservation[J]. Environmental Pollution, 2011,159: 1974-1983.

[27]LA GRECA P, LA ROSA D, MARTINICO F, et al. Agricultural and green infrastructures: the role of non-urbanised areas for eco-sustainable planning in a metropolitan region[J]. Environmental Pollution, 2011,159: 2193-2202.

[28]LA ROSA D, PRIVITERA R. Characterization of non-urbanized areas for land-use planning of agricultural and green infrastructure in urban contexts[J]. Landscape and Urban Planning, 2013, 109: 94-106.

[29]MANDER U, UUEMAA E. Landscape assessment for sustainable planning[J]. Ecological Indicators, 2010,10: 1-3.

[30]MCHALE M R, MC PHERSON E G, BURKE I C. The potential of urban tree plantings to be cost effective in carbon credit markets[J]. Urban Forestry&Urban Greening, 2007,6: 49-60.

[31]MUNOZ F. Lock living: urban sprawl in mediterranean cities[J]. Cities,2003,6:381-385.

[32]NOWAK D J, CRANE D E. Carbon storage and sequestration by urban trees in the USA[J]. Environmental Pollution, 2002,116: 381-389.

[33]NOWAK D J, CRANE D E, STEVENS J C. Air pollution removal by urban trees and shrubs in the United States[J]. Urban Forestry&Urban Greening, 2006,4: 115-123.

[34]OLSON R K, LYSON T A. Under the blade: the conversion of agricultural landscapes[M]. Boulder: Westview Press, 1999.

[35]RASIO R. I Suoli[M]. Milano: Ersal, 1999.

[36]SHIN D H, LEE K S. Use of remote sensing and geographical information system to estimate green space temperature change as a result of urban expansion[J]. Landscape and Ecological Engineering,2005,1: 169-176.

[37]SWIFT M J, IZAC A M N, VAN NOORDWIJK M. Biodiversity and ecosystem services in agricultural landscapes-are we asking the right questions? [J]. Agriculture, Ecosystems and Environment, 2004,104: 113-134.

[38]TARRANT M C, CORDELL H K. Amenity values of public and private forests: examining the value-attitude relationship[J]. Journal of Environmental Management,2002,30:692-703.

[39]THOMPSON C W. Urban open space in the 21st century[J]. Landscape and Urban Planning, 2002,60: 59-72.

[40]VEJRE H, SØNDERGAARD J, JELLESMARK F, et al. Demonstrating the importance of intangible ecosystem services from peri-urban landscapes[J]. Ecological Complexity, 2010,7: 338-348.

[41]VOULIGNY E, DOMON G, RUIZ J. An assessment of ordinary landscapes by an expert and by its residents: landscape values in areas of intensive agricultural use[J]. Land-use Policy, 2009, 26: 890-900.

[42]YANG J, MCBRIDE J, ZHOU J,et al. The urban forest in Beijing and its role in air pollution reduction[J]. Urban Forestry&Urban Greening, 2005,3: 65-78.

[43]ZOLNIK E J. The costs of sprawl for private-vehicle commuters[J]. Journal of Transport Geography, 2012,20: 23-30.

The Spatial Plan of the Metropolitan Region of Barcelona

Joan López Redondo
Universitat Autònoma de Barcelona
juan. lopez@ uab. cat

ABSTRACT: In April 2010 the Catalan Government approved the Spatial Plan of the Metropolitan Region of Barcelona. It was the first spatial supra-local plan ever adopted for a region of 5 million inhabitants, 3,200 km^2 and site of one of the most important industrial areas in Europe.

The Plan had to meet the negative impacts that land use patterns had generated in previous years as well as the foreseen demands for new housing and spaces for economic activity. According to the criteria of the Catalan Spatial Planning Programme, it was necessary to give an answer that was simultaneously able to preserve the open spaces system, ensure economic competitiveness and facilitate social cohesion.

In order to achieve this objective, the Plan proposals were structured around three main axes: the concentration of future urban developments in a system of nodes; the consideration of open spaces as a system at the same level as human settlements and transport infrastructure, on which specific planning proposals could be made; and the design of a transport infrastructure oriented to avoid overloading of roads and traffic mix, to increase system connectivity by superimposing an orthogonal structure to the existing radial one, and to deepen the capillarity of the network.

Key Words: Barcelona, spatial planning, metropolitan region, urban sprawl, land use, Catalonia

The Metropolitan Region of Barcelona (MRB) is one of the seven regions into which Catalonia is divided for spatial planning purposes (Figure 1). The MRB's area is 323,553 hectares, and its population in 2012 was 5,051,502 inhabitants (Figure 2). Over 1.6 million of these inhabitants live in the municipality of Barcelona itself (100 square km), whilst another 1.6 million are spread out among the 35 municipalities around it. Together, these 36 municipalities form the Barcelona Metropolitan Area (AMB), an administrative body since 2011.

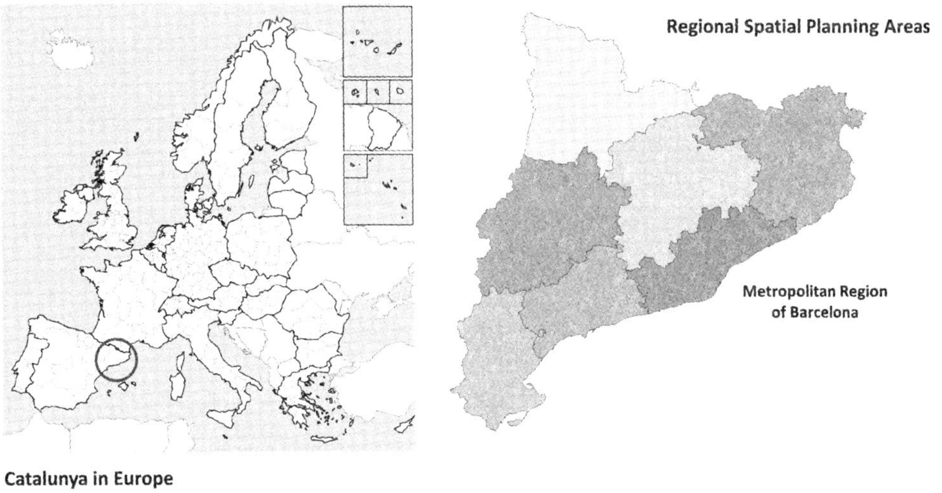

Figure 1　Soure of basemap of Europe: ESPON Hyperatlas

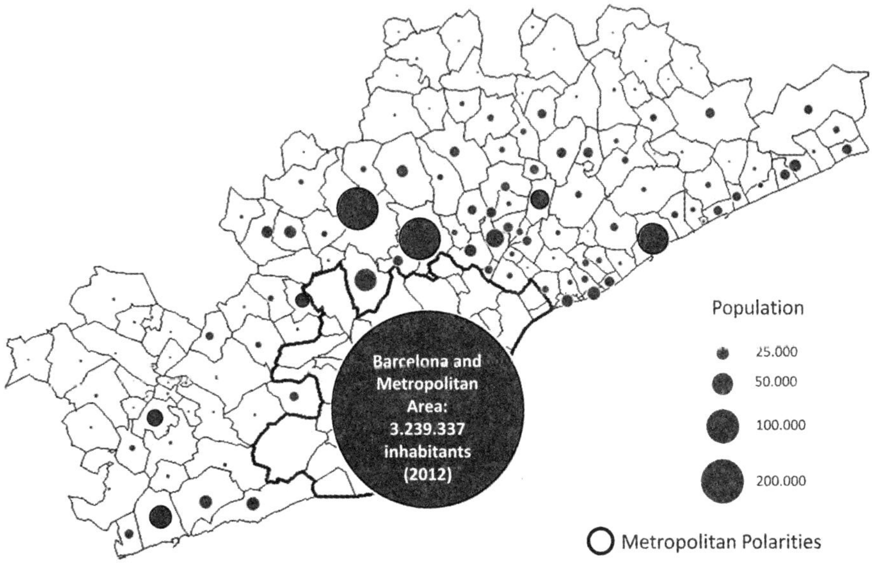

Figure 2　Metropolitan region of Barcelona

Despite the fact that most of the population of the MRB (64%) is concentrated in this small central area (636 km^2), there are also some important metropolitan polarities distributed around the region. Thus, from east to west, mid-sized cities like Mataró (125,000 inhabitants), Granollers (60, 000), Sabadell (210, 000), Terrassa (215,000), Martorell (30,000), Vilafranca (40,000) and Vilanova (65,000) play an important role in the articulation of the metropolitan region and in the way it operates as an integrated territory.

1 EVOLUTION OF THE METROPOLITAN SYSTEM

1.1 GENERAL TRENDS

Over the last sixty years the MRB has experienced a complex settlement process. In this process, three stages can be roughly distinguished, corresponding to significantly different intensities and occupancy patterns.

During the first stage, ranging from the early 1950s to the late 1970s, the AMB basically doubled its population thanks to very strong immigration flows arriving largely from the rest of Spain. Thus the population grew significantly from 1,966,291 inhabitants in 1950 to 4,238,876 in 1981.

During this period the population tended to settle near large centres of activity, that is, the city of Barcelona and the municipalities next to it, but also in the larger cities placed in the second metropolitan ring. These nuclei concentrated major housing supply operations and, because of their proximity to activity centres, produced a form of growth that was highly spatially concentrated. In the context of the poor accessibility that characterised the metropolitan region during this period, people had to locate their residence as close as possible to their workplace.

In the second stage, beginning in the late 1970s, the widespread economic crisis slowed the arrival of those important migratory flows largely due to reduced employment opportunities. A period of demographic stagnation followed, extending over nearly two decades where the metropolitan population was stable with around 4.2 million inhabitants.

Yet various centrifugal forces that accumulated over the previous stage, such as the saturation of urban areas that experienced high growth, differences in the price of housing or the appreciation of other elements of the environment, took advantage of progressive improvements in accessibility due to a expanding road network and an increased use of cars. Thus, much of the population relocated over increasingly larger areas of the metropolitan territory. The difference in the price of housing, not only between areas of origin and destination but also among destinations themselves, on the other hand, contributed to configure a heterogeneous territory with regard to its social fabric.

New dynamics emerge in the third and final stage that starts in the late 1990s, bringing new elements of complexity that coexist with the previous phases. These dynamics are clearly characterised by the restart of population growth, which enables the Barcelona metropolitan population to overcome the period of stagnation it was immersed in for almost two decades. With annual growth rates of 1.9 %, that between 2001 and 2006 exceeded even those of the 1970s (1.7 %), the metropolitan population reached 5,012,961 inhabitants in 2010. Similar to the first stage, population growth relied heavily on the sudden arrival of significant migratory contingents but unlike then, these flows were comprised primarily of foreign populations of mostly non-EU citizens. Beyond their composition, however, the common feature of the two processes lies in their location patterns, always tending to concentrate in the major cities and urban areas. This fact, coupled with the movements of population from cities to less dense areas of the first and, especially, the second metropolitan ring, enables major urban centres to reverse the negative demographic trends that characterised them for years.

Regarding economic activity, despite the moderate growth rates of the metropolitan economy in comparison with the rest of Catalonia, the Region's production accounts for over two thirds (68.9%) of the total. The metropolitan economic structure has traditionally been based on a robust industrial sector, with a significant manufacturing component supported by fairly diversified services and a primary sector that, despite its structure, has been able to adapt to the new economic conditions and the needs required by the contemporary industrial period.

In a process initiated decades ago, however, the metropolitan economy, like most Western economies, has shown a marked trend towards a growth in the service sector: in 2001, one quarter (26.3 %) of the jobs were industrial, significantly lower than figures from just a decade earlier (37.5 % in 1991).

From the spatial point of view, a profound process of activity relocation, both internationally and within the region itself, must be highlighted. Aside from allowing an increase in exports and imports, the increasing internationalisation of the Catalan economy has attracted foreign companies investments as well as Catalan capital abroad, giving way to part of the production process, in particular industrial production, being transferred overseas.

Domestically, the relocation has basically consisted of the exodus of those activities that have difficulty cohabitating with other urban actors from dense urban areas, because they are big land consumers or simply because of their accessibility needs. Similar to population distribution patterns, activities, largely industrial and logistical, have taken advantage of the new accessibility conditions to move where land supply is more plentiful and cheap and where the operating efficiency of the entire production system can be maximised. As a result, those activities have had a tendency to extend over the whole region, so that at present it is distributed in a more homogenous way.

1.2 SPECIFIC PATTERNS

Over the last few decades, both population and productive activities have experienced a double process of growth and redistribution at a metropolitan scale. Not only location has had a significant regional impact on this redistribution, but also the shape or the patterns of urbanisation followed by new residential and business developments have played a role.

Looking at the residential land use, it is important to read this urbanisation process from both the number of new housing units built as well as-and above all-their typology. It is thus true that this stage of demographic stability, between 1980 and the turn of the century, enabled a correction of many deficits accumulated over the preceding decades in terms of housing, services and facilities. In this sense, the de-densification of the areas that had experienced more intense growth should not be interpreted as negative but rather as the satisfaction of a latent demand. Furthermore the construction of the housing units, required to reduce excessive occupancy rates, enabling families to extend into different dwellings or to dignify the living conditions of their inhabitants, inevitably brought with it the need to occupy new land.

The way in which this de-densification took place leads to describe this process as extensive in terms of land development. Many municipalities, especially Barcelona and those of the first metropolitan ring, based their interventions on new multi-family housing types and higher densities, in order to make the most of the limited land reserves available or because they considered these high densities more efficient from the standpoint of urban functionality. However, many other municipalities, sometimes conditioned by the urban planning they inherited, tended to promote low density housing developments.

At the same time, this process of the metropolitan territory's recent urbanisation has been characterised by dispersion. It tends to occupy new spaces, often small in dimension, away from existing urban areas and the main transport network, both road and, above all, railway. That is, not only is land less intensively occupied, but the de-densification process is also accompanied by a subsequent process of dispersion where in new developed areas break the former compactness traditionally associated to urban fabric (Figure 3).

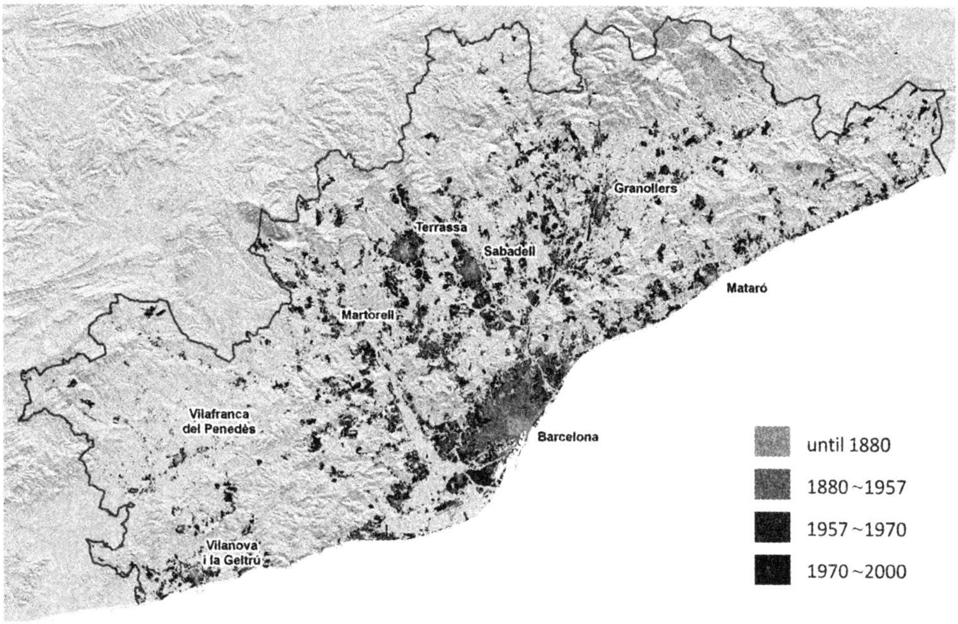

Figure 3 Evolution of land developed in the metropolitan region of Barcelona

Similar to residential development, the process of relocating industrial activities from major cities to the periphery described previously is largely based on the model of industrial zones from the 1960s. To some extent it was positive, since these areas have greater availability of space, are designed specifically for productive activity and are usually segregated from the existing urban fabric. On the other hand, the proliferation of industrial land supply in most municipalities has consolidated this trend and led to a scenario of multiple scattering of industrial areas, sometimes with very small dimensions.

In this context, moreover, the process of peripheralisation of services often relied on the occupancy of these new industrial zones by tertiary activities such as trade or leisure. This led to a perversion of the original purpose these areas were created for, so that industry currently occupies about half of the surface of the metropolitan industrial zones, while logistics have gradually increased their share.

To sum up, it is not only the physical form, dominated by scattering, that characterised the process of relocation of the activities on the metropolitan territory, but also the occupancy of these new spaces for uses they were not initially intended for.

On the other hand, next to this dispersion and peripheralisation pattern of service activities primarily addressed to the population and, therefore, eminently urban, there are also numerous initiatives based on the creation of spaces integrated into the existing urban fabric during recent years, where economic activities coexist with residential and urban space.

1.3 ANALISYS OF IMPACT AND CAUSES

Physically, these processes of urban development resulted in a very high consumption of land, so that the area transformed in the metropolitan region increased 36.5%, from 56,314 hectares in 1977 to 76,884 hectares in 2000. The lower density led to a much higher per capita urbanisation of land than had occurred before.

The negative impacts of this urbanisation process, however, are not limited to an absolute quantification of land taken up, but also manifest through indirect effects. Its spatial patterns led to a new distribution of activities: what once was mixed now is separated; what once was close now is far. And, as a consequence of this functional specialisation and the gradual distancing and segregation of urban pieces, the population is forced to move farther and farther to satisfy that need formerly satisfied locally.

Dispersion and specialisation lead to increased mobility, which has associated economic, time, pollution and accident costs for people. And, moreover, the need to travel in itself requires more and larger transport infrastructure that, due to its linear structure, entails significant land fragmentation. In this way, the population, as well as productive activities and the environment, will suffer some negative impacts that initially had not been considered (Figure 4).

One of the main causes of the negative impacts produced by land use patterns is the lack of a spatial planning figure at a supra-local level. The primary causes are obviously diverse and, in most cases, far from the spatial planning sphere: a significant population growth, a parallel rise in economic activity and an increase in the standard of living of this population that in turn opens new possibilities of consumption and improved living conditions also in regard to the characteristics of the residence. All these elements were responsible for a greater urbanisation of land in the Metropolitan Region of Barcelona o-

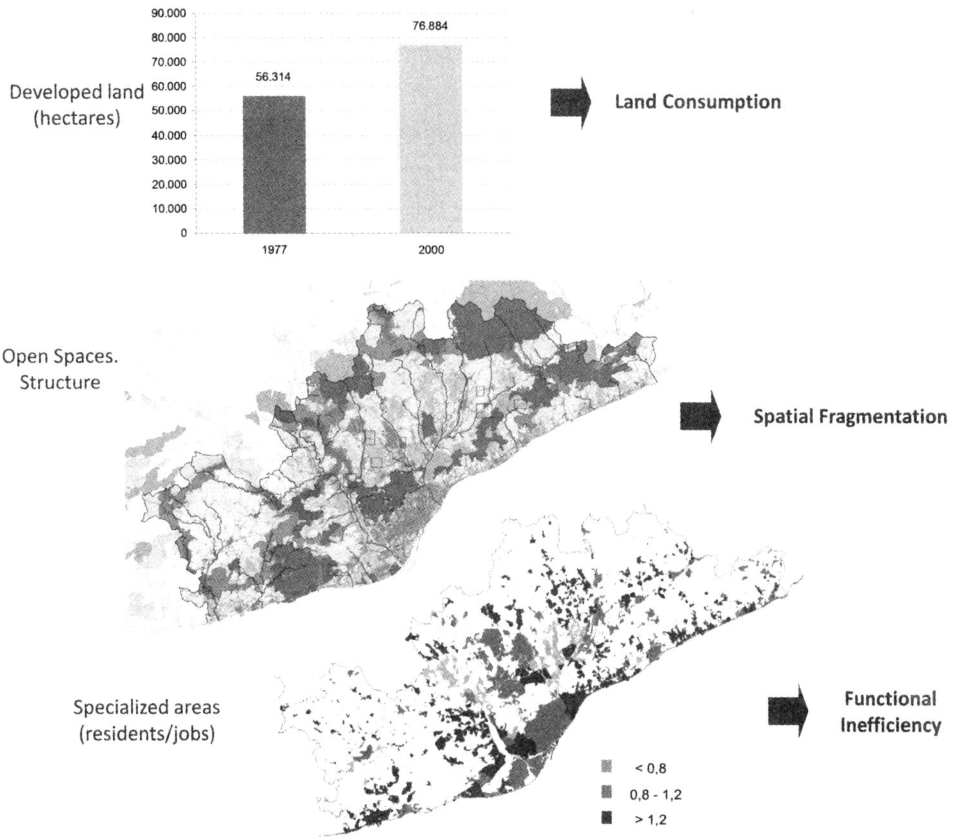

Source: Spatial Plan of the Metropolitan Region of Barcelona

Figure 4 Impacts of urbanisation in the Metropolitan Region of Barcelona

ver the last decades. Nevertheless, the fact that the spatial crystallisation of this process took one path or another depended largely on supra-local planning.

Catalonia has long had a well-developed tradition of urban planning at the municipal level. But the dynamics described above corresponded to a larger regional and metropolitan scale. It is true that there had been numerous supra-territorial planning proposals in Catalonia, but indeed none of them were approved. So, leaving aside the experience of the diluted 1995, Pla Territorial General de Catalunya, the first regional plan approved in Catalonia actually dates from 2006. In a context of Catalonia's population and economic growth during the last 60 years, in a context in which this growth is furthermore accompanied by de-densification processes and urban sprawl, the lack of spatial planning capable of coordinating new supra-local scale developments has had, as described, largely negative impacts.

For this reason, in 2004 the Catalan government decided to promote a Spatial Planning Programme, with the aim of providing tools to correct the dispersion, specialisation and segregation trends of supra-municipal scale urbanisation. The Spatial Plan of the Metropolitan Region of Barcelona is located in this framework programme, a Plan adopted in April 2010 (Figure 5).

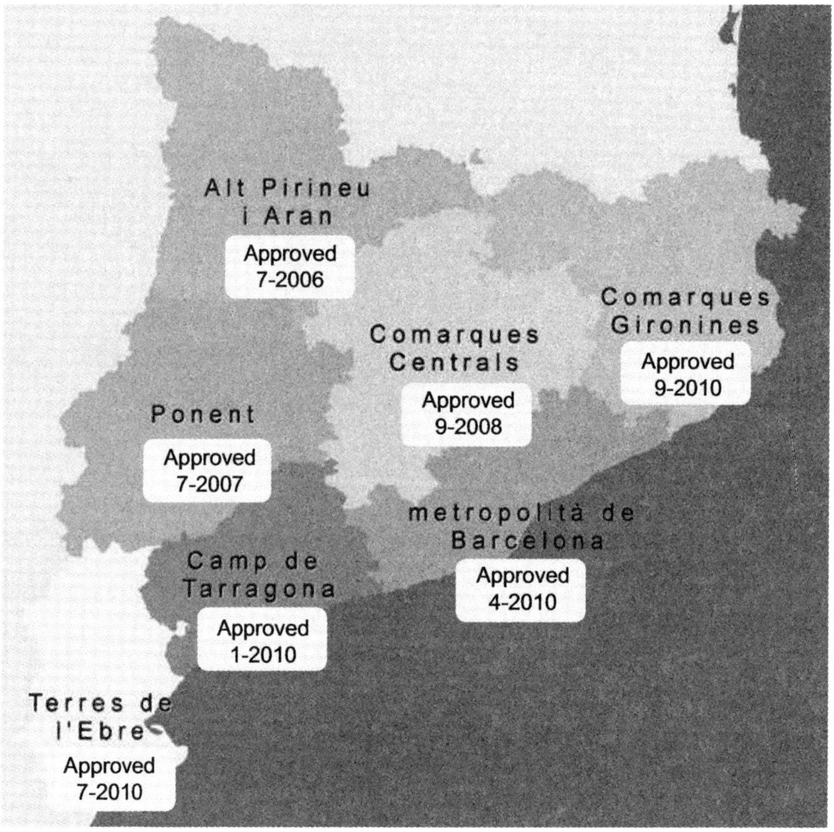

Source: Spatial Plan of the MRB.

Figure 5 Catalan Regional Spatial Plans (2006 ~ 2010)

2 THE MAIN CHALLENGES THAT THE MRB SPAZIAL PLAN HAD TO FACE

The Spatial Plan of the MRB had to address important spatial challenges from the outset. The population and activity scenarios developed in the Plan for the next fifteen or twenty years showed an additional increase in both population and economic activity. Thus, from the most plausible hypothesis, the population of the MRB would reach 5.4 million in 2026, which meant 940,000 inhabitants more than it had in 2001. It also foretold the creation of 650,000 additional jobs.

These foreseeable increases would result in demands of housing units and land for business. In this sense, the future needs amounted to 592,000 new housing units and the land required to accommodate the said 650,000 new jobs.

It was, therefore, targeted to meet these demands for new housing and spaces for economic activity. And considering the negative impacts that land use patterns had generated in recent years, it was necessary to give an answer that was simultaneously able to preserve the open space system, ensure economic competitiveness and facilitate social cohesion.

3 URBAN STRATEGIES AND CONCEPT ELEMENTS

3.1 NODAL SYSTEM

Trying to devise a formula that takes into account all these requirements, the Spatial Plan started from a basic concept: the concentration of future urban developments in a system of nodes. That is, it was necessary to locate new housing needs and economic activity, but this had to be done by minimising land consumption and the impact on open spaces, and by optimising the existing infrastructure. In this sense, the nodal system allowed a proportioned concentration of future growth by means of two basic strategies.

First, the most important developments should be concentrated in a few urban centres. Unlike other widely known experiences, however, in the case of the MRB it seemed more appropriate to take advantage of the already existing urban system. This system is articulated from a powerful centre of more than 3 million people formed by Barcelona and its conurbation. Around this core area, there are a series of intermediate cities between 30,000 and 200,000 people, structuring centres of the surrounding territories and able to articulate them with the rest of the metropolitan region. Some of these centres, the biggest ones, experienced an enlargement of their urban core in the nineteenth century. This enlargement, as it occurred in Barcelona as well as in cities like Sabadell, Terrassa or Mataró, allowed these centres to multiply their population and move from being small cities to metropolitan centres. Others, however, maintained their development potential as centres articulated with nearby areas due to their location and type of relations with the rest of the metropolitan system. The first proposal of the MRB's Spatial Plan was, therefore, to take advantage of these existing potentialities and concentrate there in much of the future demand for land for housing and economic activity. Their ability to structure neighbouring territories and their infrastructure provision would be, thus, well accompanied by the necessary critical mass to develop their true function as metropolitan pole (Figure 6).

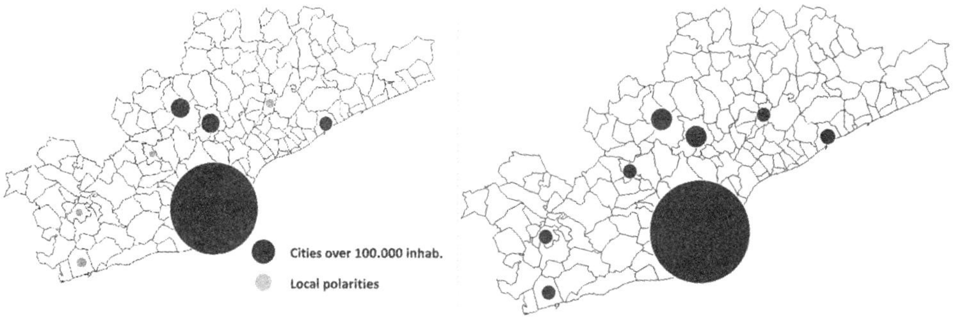

Figure 6 From local polarities to metropolitan centres

Second, once these nodes were granted the largest stake of future growth, the remaining towns would host the other demands for land required for housing and other activities according to their characteristics, size and location. With this purpose in mind each town and village was designated a specific growth strategy (Figure 7). Thus, not only was the isolated and segregated proliferation of new urban developments prevented in the existing urban areas, but it was also advocated that these developments were proportionate to the urban areas where they were taking place.

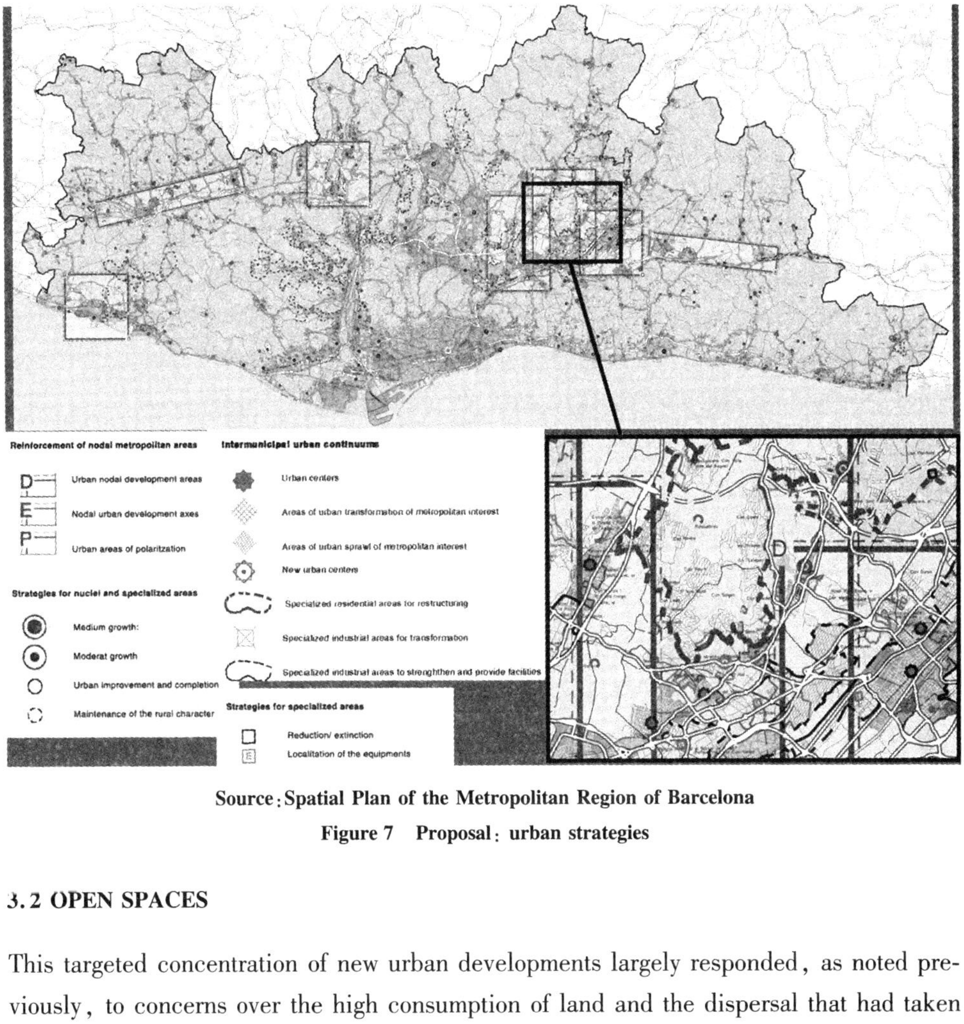

Source: Spatial Plan of the Metropolitan Region of Barcelona

Figure 7 Proposal: urban strategies

3.2 OPEN SPACES

This targeted concentration of new urban developments largely responded, as noted previously, to concerns over the high consumption of land and the dispersal that had taken place over the preceding decades. During this period, the extension of metropolitan open spaces was strongly reduced, as was the spatial continuity that enabled their functionality. For this reason, open spaces were considered by the Spatial Plan of the MRB as a system, at the same level as human settlements and transport infrastructure, on which specific planning proposals could be made. This was, in a sense, completely new in Catalonia: spatial planning should not focus exclusively on urban development and infrastructure while open spaces were considered "the remainder", or the area at the service of future developments. From now on, open spaces would play an active role in planning, equally with the other systems considered.

For this purpose, the Plan was based on the consideration of the entire metropolitan land already having some level of protection, from local to international, and included it in the so-called open spaces system. This entire area, 242,000 hectares (three quarters of the metropolitan region), would be considered part of the open spaces system and, as such, it would have some protection.

Within this system, every space received a specific level of protection in function of both its intrinsic value and role in the whole system. That is, the greatest protection was not only given to those areas with special interest for their fauna, agriculture or forestry values, but also to those spaces that connected these areas and, thus, allowed its functioning as a system. These areas of interest (205,000 hectares) received the level of protection termed "special", which excluded them from any new development.

The rest of the open spaces system (nearly 40,000 hectares) received the protection designated as "preventive". It identified land that was not developed but was susceptible to host the developments considered by the Plan itself, with a margin of demarcation sufficient to meet the land needs for new developments wherever local plans consider the most appropriate, but without setting a specific site. That is, allowing new developments where they have less negative impacts and, after they had taken place, keeping the rest of land as part of the open space system (Figure 8).

3.3 TRANSPORT INFRASTRUCTURE

Finally, proposals for transport infrastructure in the MRB's Spatial Plan were in line with those of the two previous systems: promoting a nodal urban structure and preserving as much as possible of the open space system. The achievement of this objective required first overcoming three determinants of the existing configuration of the transport infrastructure, both in matters concerning the road and the railway networks. These determinants are the result of the networks' historical development and the various functions they have acquired in the process of the development of the urban system.

The first major determinant was the absence of a transportation network conveniently segregated and organised by hierarchy in function of its uses. In terms of the road system this resulted in a mixture of traffic in the interior of the MRB with very diverse journey objectives, duration and distances travelled. The most obvious example of this is probably the so-called by-pass "Y", linking the main crossing of the Pyrenees to the Mediterranean axis and the interior of the Iberian Peninsula, which passes through the interior of the MRB. Without being a negative characteristic in itself, the absence of lower-level

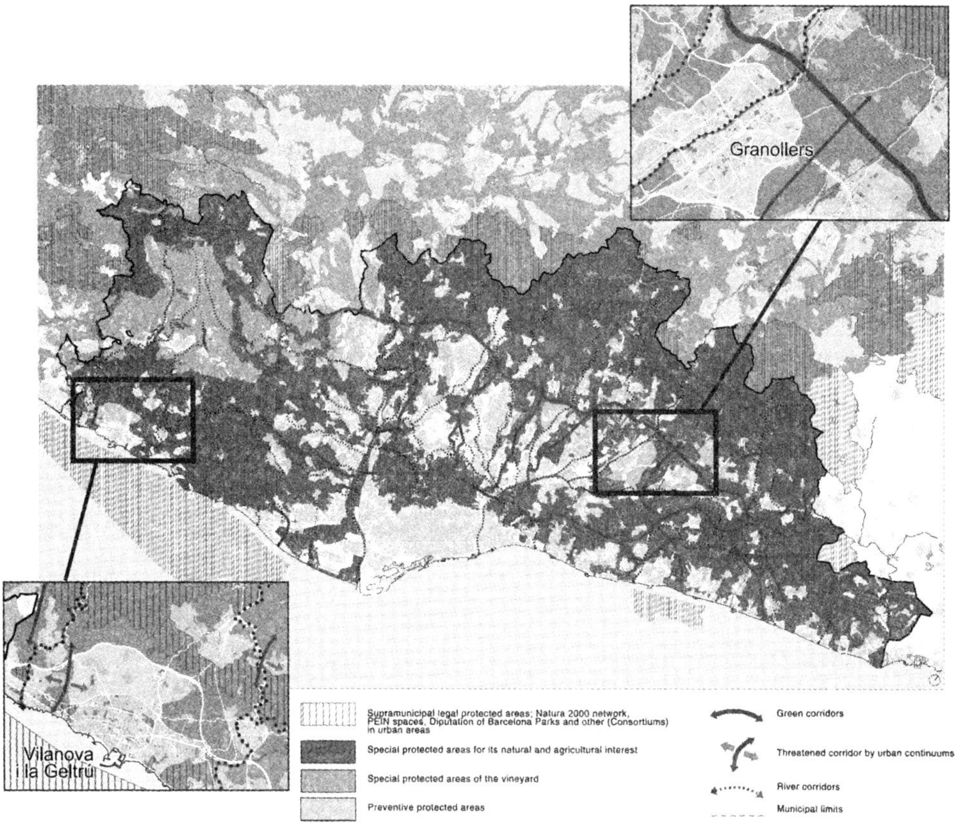

Source: Spatial Plan of the Metropolitan Region of Barcelona.

Figure 8 Proposal: open spaces system

roads designed to serve local traffic means that the same artery used to communicate Europe to the rest of Spain or the African continent is used to travel between neighbouring municipalities within the metropolitan region as well. Clearly, journeys of such diverse natures require roads with diverse characteristics in order to function in an efficient manner. The "Y" is also observable in the rail network today, which acts to limit by-pass rail traffic, especially those trains carrying goods or merchandise.

The second constraint present in the transport infrastructure network of the metropolitan region was its eminently radial structure. The robustness of this radial structure derives from the great weight of the city of Barcelona and its metropolitan area as the centre of the whole territory of Catalonia. Both the formation of the motorways and rail networks (the so-called "Catalan eight") show a strong concentration in the metropolitan centre. As a result, there is not only a disregard for the connections between the cities of the metropolitan periphery with each other, which requires the passage through the centre of the system in virtually all cases, but an overload of this centre due, to a great extent, to these unnecessary trips.

The third major determinant of current transport networks in the MRB was its lack of capillarity, this is, the lack of an intermediate level. With regard to the road network, it is clear that there is an acceptable offer of motorways that allows connection between major metropolitan nodes. There is also a good supply of local roads that connect small and medium towns with each other. But the system lacks an intermediate network of roads that allow an easy transition between motorways and local roads. In the case of railways, the situation is similar: efficient urban transport networks like the metro and tram lines have joined a powerful array of connections to the larger scale high-speed train. Meanwhile, the intermediate level of service, that efficiently connects the various parts of the metropolitan territory, maintains a service that can clearly be improved.

Given these determinants, the Spatial Plan proposals were targeted to avoid overloading of roads and traffic mix, to increase system connectivity by superimposing an orthogonal structure to the existing radial one, and to deepen the capillarity of the network (Figure 9 and Figure 10). It should be added that, in all cases, the actions prioritised the rail network over the road, because of its bigger capacity to serve and articulate the metropolitan region territory.

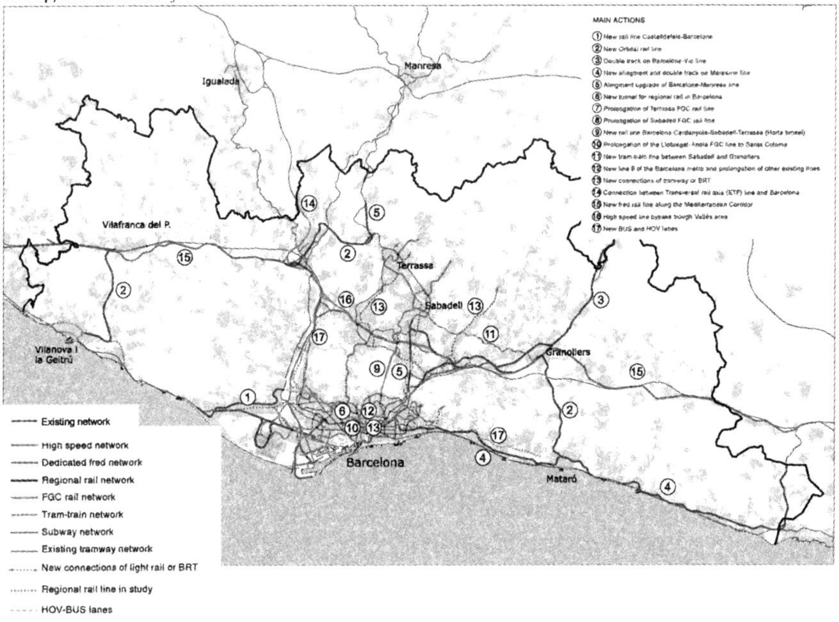

Figure 9 Proposal: railway network

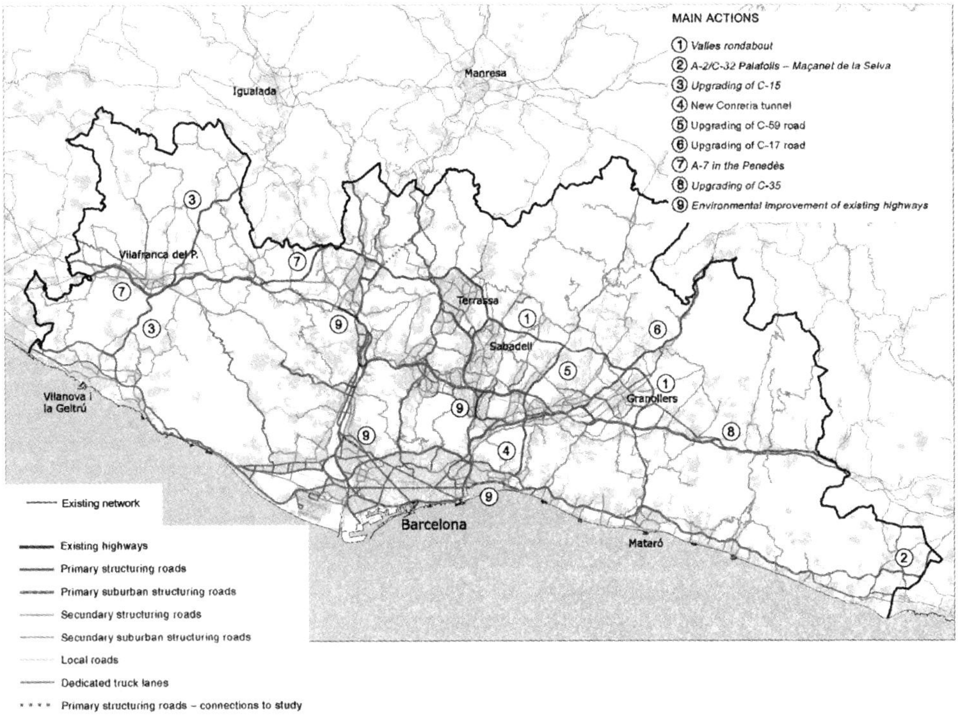

Source: Spatial Plan of the Metropolitan Region of Barcelona.

Figure 10 Proposal: road network

4 CONCLUSIONS

The spatial plan of the Metropolitan Region of Barcelona (MRB) was adopted, as has been said, in April 2010. The plan sought to answer a series of challenges that arose over the preceding decades and had not received an effective response from territorial planning at the supra-local scale. These challenges, which threated the environmental sustainability, social cohesion and economic competitiveness of the metropolitan territory, had already been recognised not only by those responsible for land management, but also by most agents involved: municipalities, civil society organisations and a large stake of the population.

This recognition greatly facilitated relationships characterised by extensive social participation. Indeed, the process of preparing the plan was designed using a thorough participatory mechanism where local authorities and other agents had the opportunity to make contributions and to express their agreement or disagreement with the proposals that were created. But in addition, the plan itself was impregnated with that open and participatory spirit, and the level of dialogue with a range of partners was far beyond that established by purely formal requirements. For this reason, and beyond the level of success of the proposals, the Spatial Metropolitan Plan (as well as the rest of regional plans approved in Catalonia at that time) had a high degree of social acceptance at the time of its approval.

Thus, the MRB finally had an approved, appropriate and accepted plan. However, these three requirements are not sufficient to ensure the full achievement of the objectives set by any spatial plan, as territorial planning activities involve several factors that cannot be met through the planning instrument alone. These factors cannot be taken on directly through the concrete proposals of a spatial plan but, however, will decisively influence the extent to which it is achieved. In the case of the MRB, considering its characteristics and the particular timeframe for the plan's approval, three elements are particularly relevant.

The first of these elements has to do with the limitation of the planning activity itself. The spatial plan expresses a particular territorial model through its proposals. In this sense, it creates and promotes the spatial conditions for future developments to follow certain rules considered desirable. But it cannot directly affect the behaviour of economic activities when choosing a particular location, or of the population that may prefer a specific residential typology or follow certain mobility patterns. Spatial planning is limited because the extent to which proposals are implemented will depend on its ability to maximise the satisfaction of various demands, and these demands are not always guided by the same macro-criteria as those used for spatial planning. In the MRB, where over the past few decades there has been an unprecedented increase in income levels and where present generations have had a chance to satisfy their preferences at a level that their parents never had, the correspondence between individual and collective interests have not always coincided. Thus, for example, the aspiration for a certain residential typology that, in aggregate, adversely affects variables such as ecological balance or mobility efficiency is still common today among the population. In this sense, the spatial plan may favour certain models of urban development, but the successfulness of their implementation will depend largely on other societal values and preferences far beyond the spatial sphere.

The second major constraint is the time horizon required for spatial planning. Spatial plans, as is known, make long term proposals, not only because of the durability of their impact, but also for the duration of the implementation process. The period spanning from the adoption of any spatial plan to its further specification through derived planning instruments, the beginning of its development, the completion of urbanisation and construction processes and, finally, the effective occupation by people or given economic activities may last 5, 10, 15, 20 or 30 years. In this sense, the error of requiring immediate responses from spatial plans to circumstantial demands or needs arising from the economic cycle can produce some disappointment regarding their usefulness to meet social demands. In the case of the MRB, the approval of the spatial plan, taking place just as the current economic crisis began, sometimes entailed this demand for responses to needs including economic stimulation or increased competitiveness and, consequently, various actors were disappointed when they encountered rigidity in this regard. Understanding spatial planning timeframes and being able to gather enough perspective to evaluate the benefits of the spatial plan in the long term are thus some of the main factors for its survival.

The third conditioning element is the need to ensure the continuity of proposals beyond the period in which the government that drives them remains in power. So, since a spatial plan, as stated above, is the expression of a particular territorial model, it is not free from a particular ideology. On the other hand, the long spatial planning implementation process means that its full development rarely sees the same political configuration remain in power. In this sense, if the new leaders do not share the criteria underlying the current plan, it is expected that one of three common solutions will be adopted: its annulment (if they do not support spatial planning), its replacement with a different plan (if they continue believing in spatial planning, but under other criteria) or the gradual abandonment of its principles, guidelines, standards and mechanisms for monitoring (if they do not have a clearly defined position on spatial planning or, even having it, prefer not to express it).

In the case of the MRB spatial plan, as in the rest of spatial plans approved between 2006 and 2010 in Catalunya, its evolution in this regard remains unclear. At the end of 2010 the ruling coalition that had been in power between 1980 and 2003 won a majority in the Catalan government elections; during their electoral campaign they indicated a desire to revoke the plans. During these first years in office this initial will appears to have changed, without becoming either over-enthusiastic in developing the content of regional plans nor monitoring them. In any case, it has lost much of its momentum.

Overcoming these determining elements, as mentioned, goes beyond the figure of the spatial plan. An extension of the "territorial culture" is required to understand the limitations of spatial planning and to give it sufficient consideration towards ensuring the full achievement of its objectives. In this sense, even if carrying out the MRB spatial plan was discarded due to one of these reasons, the mere approval and the development process that preceded it would have already achieved this essential mission: to increase the culture of spatial planning and awareness of the benefits of its use.

REFERENCES

[1] BARBA J, MERCADÉ M. Les urbanitzacions a la província de Barcelona: localització i característiques dels sistemes de baixa densitat residencial[M]. Barcelona: Diputació de Barcelona, 2006.

[2] CARRERASC (dir.). Atlas comercial de Barcelona[M]. Barcelona : Ajuntament de Barcelona, 2003.

[3] FONT A, LLOP C, IVILANOVA J M. La construcció del territori metropolità. Morfogènesi de la regió urbana de Barcelona[M]. Barcelona: Mancomunitat de municipis de l'àrea metropolitana de Barcelona, 1999.

[4] INDOVINA F, (coord.). La ciudad de baja densidad: lógicas, gestión y contención[M]. Barcelona: Diputació de Barcelona, 2007.

[5] LÓPEZ J. La mobilitat de les persones a la regió metropolitana de Barcelona, en papers[J]. Regió Metropolitana de Barcelona, 2003, 38: 11-27.

[6] LÓPEZ J. El procés de segregació social als municipis de la regió metropolitana de Barcelona: la població immigrada[M]. Barcelona: Institut d'Estudis Territorials, 2007a.

[7] LÓPEZ J. El procés de segregació social als municipis de la regió metropolitana de Barcelona: la gent gran[M]. Barcelona: Institut d'Estudis Territorials, 2007b.

[8] LÓPEZ J. Les migracions residencials extramunicipals a la regió metropolitana de Barcelona en el període 2000-2005[M]. Barcelona: Institut d'Estudis Territorials, 2007c.

[9] LÓPEZ J. Global challenges and strategic spatial planning in the region of Catalonia [C]// VANAUGTGAERDEN L, (coord.). Polycentric regions facing global challenges: A role of strategic spatial planning. Brussels: Flemish Government, 2011, 11.

[10] LÓPEZ J, REY A. Inmigración y segregación en las áreas metropolitanas españolas: la distribución territorial de la población no europea en el período 2001-2006[J]. Institut d'Estudis Territorials, 2008: 35.

[11] MONÉS M A, CARRERA J M. La Barcelona metropolitana els propers 20 anys: Prospectiva del mercat de treball, demografia i habitatge[M]. Barcelona: Ajuntament de Barcelona, 2003.

[12] NELLO O. Ciutat de ciutats[M]. Barcelona: Empúries, 2010.

[13] NELLO O. El planeamiento territorial en Cataluña[J]. Cuadernos Geográficos, 2010, 2: 47.

[14] NELLO O. Estrategias para la contención y gestión de las urbanizaciones de baja densidad en Cataluña, in Ciudad y territorio[J]. Estudios territoriales, 2011, 43: 167.

[15] SERRA J. Dinàmiques metropolitanes a l'àrea i regió de Barcelona[M]. Barcelona: Mancomunitat

de Municipis de l'Àrea Metropolitana de Barcelona, 1994.

[16] SERRA J. El territori metropolità de Barcelona. Dades bàsiques, evolució recent i perspectives [M]. Barcelona: Mancomunitat de Municipis de l'Àrea Metropolitana de Barcelona, 2003.

[17] VALLÈS J M, NELLO O. De ciutat a metròpoli: Notes per a una lectura del Pla Estratègic Barcelona 2000 des d'una perspectiva metropolitana [M]. Ajuntament de Barcelona: Barcelona economia, núm. 18, 1993.

Mobility and Urban form:
The Case of the Metropolitan Area of Barcelona

Carles Donat
Institut d'Estudis Regionals i Metropolitans de Barcelona. Universitat Autònoma de Barcelona.
carles. donat@ uab. cat

ABSTRACT: This paper analyses the relationship between urban form and urban mobility taking as a reference the metropolitan area of Barcelona. The chosen approach focuses on the urban change from 1980s characterised by two main population dynamics: decentralization and dispersal. The paper shows how urban mobility has been increased, both in terms of displacements number and distance. Furthermore, due to the growth of the population living in low-density dispersed settlements, motorized transport, especially the private car, tend to gain weight in the overall mobility.

Key Words: urban form, urban mobility, metropolitan area of Barcelona

1 INTRODUCTION

This short paper is part of a publication which discusses about the possible future of the city, focusing on the balance between city and environment, an equilibrium in which there is the future of the planet because more than half of the world population lives in cities and it is expected that within fifteen years it will increase until the sixty will be about sixty percent.

This paper will contribute to this debate taking as a reference the metropolitan area of Barcelona, focusing the discussion on the balance between the city and the environment, on a single theme: urban mobility. Specifically, within this broad field of study, it will be focus on those aspects related to urban form. This text is structured in four main points. The first section introduces the main characteristics of the metropolitan area of Barcelona. The second one proceeds to analyse the major trends in urbanization process, highlighting the changes that occurred in the urban form. In the third section there are shown the major trends in urban mobility and its close relationship to the urbanization processes. Finally some brief conclusions will be presented.

2 THE METROPOLITAN AREA OF BARCELONA

The metropolitan area of Barcelona is located in the northeast of the Iberian Peninsula, in the region of Catalonia. Its extension is near 3,200 km^2 with a population that exceeds the five million of inhabitants. The metropolitan area of Barcelona, together with Madrid and Lisbon is on top of the Iberian metropolitan system. Furthermore in the European context it is among the top 10 urban agglomerations. As well as other major European cities its economy is clearly dominated by the service sector, which employs 79 % of the population. Nevertheless, Barcelona is still the main centre of the industrial production in the peninsula and the main exporter.

Like many other cities, Barcelona experienced a very intense population growth, parallel to the industrial revolutions occurred since mid-eighteenth century, than in the case of Barcelona it had its peak point in the first third of the twentieth century. From mid-fifties to mid-seventies of the twentieth century, and coincident with the end of the international isolation of the dictator of Spain, resulted in another period of intense economic and urban growth. All these steps of growth and urban concentration occurred not only in the city of Barcelona, but also in some of the surrounding cities, where the society and the landscape were also involved in the transformation according to the new forms and relations of production.

Thus, as shown in Figure 1, together with Barcelona (1,611,013 inhabitants), the important cities of Sabadell (206,949 inhabitants), Terrassa (214,406 inhabitants), and Mataró (123,367 inhabitants), but also Granollers (59,571 inhabitants), Vilafranca (38,839 inhabitants) and Vilanova (66,074 inhabitants), were conceived as intermediate poles in the urban hierarchy. Nowadays these poles, located in the second ring, are integral part of the metropolitan area of Barcelona and its polycentric character. Above this polycentric structure, articulated on a system of compact cities with a mixture of land uses (residential, production, trade, etc.), has been developed, mainly, the recent urbanization of the metropolitan area.

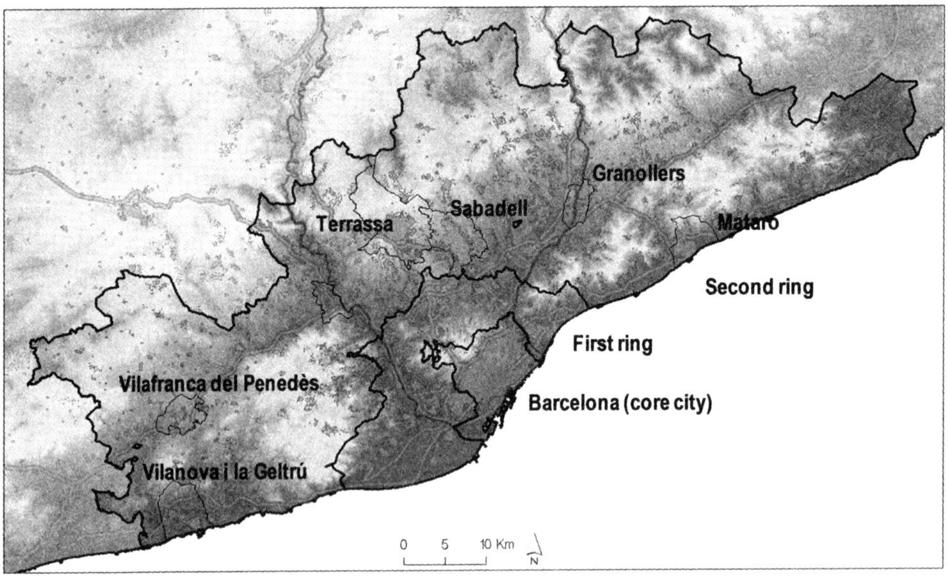

Figure 1 The territorial structure of the metropolitan area of Barcelona

However, from the 1980s there have also been new dispersed residential developments and old second dispersed houses have become main home. Additionally, there has been a proliferation of specialized areas in activities, leisure and consumption, often isolated from traditional urban settlements. Consequently, the urban sprawl pattern has gain influence against the city reasonably compacted and with a mixture of land uses, as we will see in the next section.

2 URBAN DYNAMICS AND URBAN FORM: DISPERSAL AND SPECIALIZED LOW-DENSITY URBANIZATION

As in the vast majority of contemporary European metropolis, the urbanization process in the metropolitan area of Barcelona has been characterized by the rotation of two modern stages regarding the distribution of population and activities: a first stage of concentration, that in the case of Barcelona took place until 1970s, and a second stage of decentralization, on which we focus in this section. In addition, this stage of decentralization has been accompanied by a dispersal of urbanization based on a growth of specialized low-density urban forms.

As it is well known, the end of the seventies of the twentieth century produced a set of economic and social changes of great significance. In particular, the two oil crises and the intense process of economic restructuration, along with the improvements in infrastructures and household income, just to refer the main factors, resulted in a relocation of the activities and the population. In the case of population, as shown in Figure 2, from the 1981 there was a change in trend in the metropolitan area of Barcelona: the importance of the metropolitan area in relation to the whole region of Catalonia, although moderate began to fall. Additionally, this decline in the relative weight took place in both periods of relative stability in population and periods of high demographic dynamism.

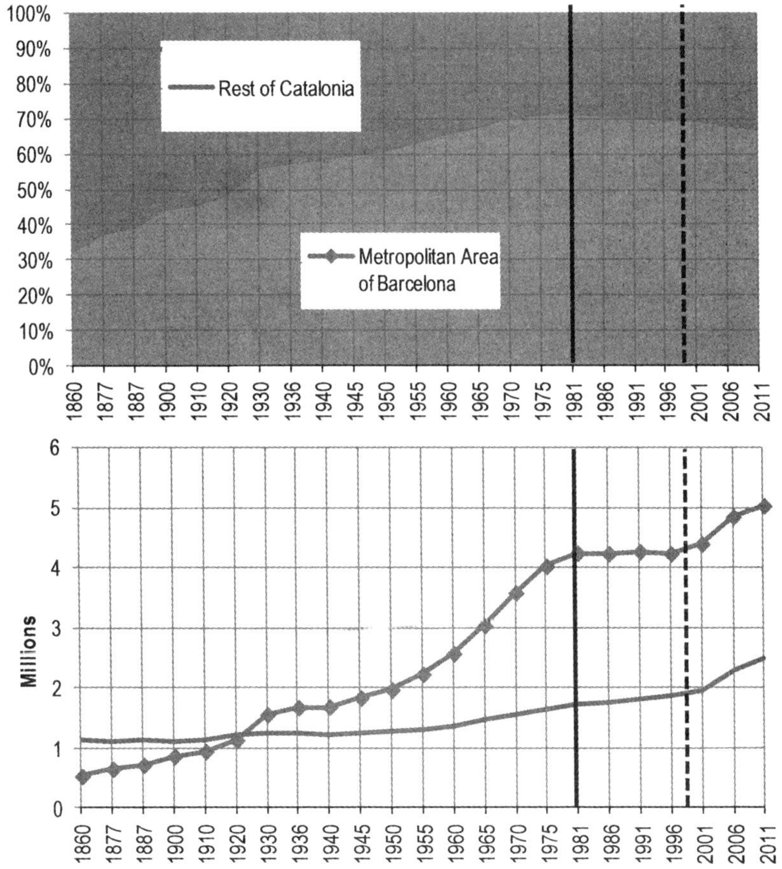

Source: Statistics Institute of Catalonia

Figure 2 Population distribution. Catalonia. 1886 ~ 2011

In fact, between 1981 and 1996, the population of the whole of Catalonia remained fairly stable at around six million. In this framework of stability, the metropolitan area experienced a decline in their relative weight, which decreased from 71 % to 69 %. And even in absolute terms, since the metropolitan area lost 10,000.

On the other hand, between 1996 and 2009, there was a widespread population growth, mainly due to international migration. And in this context, although the metropolitan area stopped losing population in absolute terms, continued to do so in relative terms, due to the fact that the rest of Catalonia grew with a higher intensity. Specifically, the weight of the metropolitan area decreased from 69 % in 1996 to 67 % in 2009. In conclusion, it can be seen clearly the route of a cycle of intense concentration to another of decentralization.

This stage in which we are immersed, common to most of the major European urban areas, has been interpreted in some cases as a symptom of decay of cities. However, a more detailed analysis of the processes just seen show that, in very good extent, this decentralization does not mean a break with the city but rather the contrary.

Figures (3a, 3b and 3c) show the residential changes produced in Catalonia in recent decades. Graphics are divided into a scheme of rings, where Barcelona is the city centre. Then it comes the rest of urban continuum, and the rest of the first metropolitan ring, then the second ring, where the industrial traditional cities are located, and finally the rest of Catalonia, that is not strictly part of the metropolitan area.

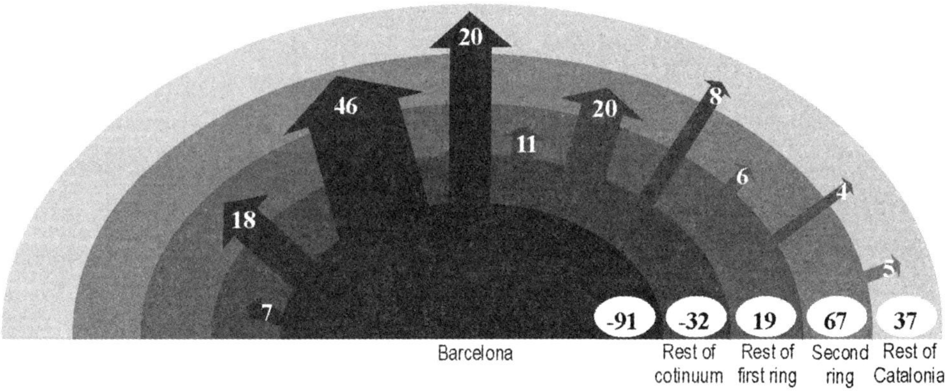

Source: Serra (1997)

Figure 3a Residential migration balances in Catalonia. 1991 ~ 1996 (thousands)

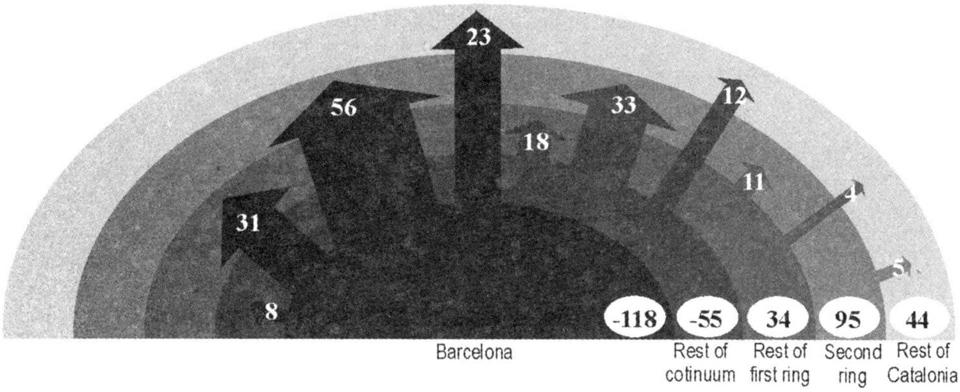

Source: INE, Estadística de Variaciones Residenciales 1997 ~ 2001

Figure 3b Residential migration balances in Catalonia. 1996 ~ 2001 (thousands)

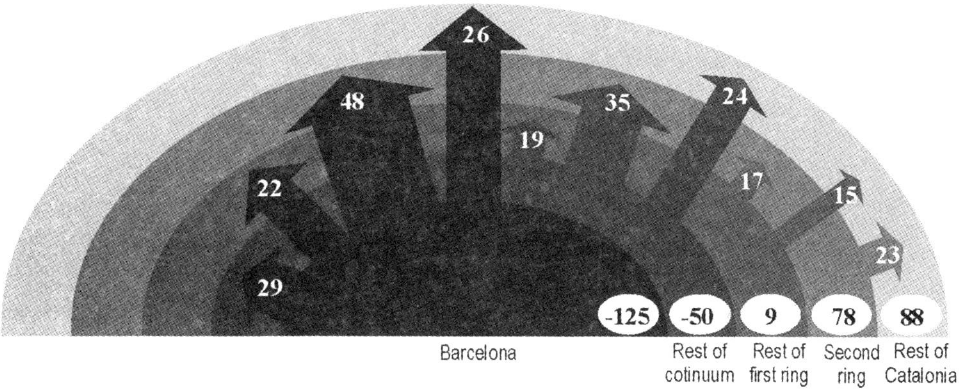

Source: INE, Estadística de Variaciones Residenciales 2002 ~ 2006

Figure 3c Residential migration balances in Catalonia. 2001 ~ 2006 (thousands)

As we can see, in all three periods, from 1991 to 1996, from 1996 to 2001 and from 2001 to 2006, residential migration balances always shows a clear predominance of centrifugal movements. As example, in the figure of the period 1991 ~ 1996, Barcelona lost 7,000 inhabitants that moved to the rest of urban continuum, 18,000 inhabitants that went to the rest of the first ring, 46,000 inhabitants that moved to the second ring, and finally 20,000 inhabitants that went to the rest of Catalonia. These centrifugal movements could also be seen in other territories. The rest of urban continuum, which as we have just seen attracting central city population, however is losing population over the outer areas: 11,000 inhabitants went to the rest of the first ring, 20,000 inhabitants moved to the rest of the second, and 8,000 inhabitants moved to the rest of Catalonia.

If we continue doing this exercise it could be seen that this centrifugal movement, this decentralization, is common to all territories. Consequently, the remotest area, the rest of Catalonia, the land that is not considered strict metropolitan area, always has a positive balance in relation to others. Observing to the following periods, analysis shows how the centrifugal residential migrations are becoming more intense. In addition, migration generate that the outermost territory, the rest of Catalonia, grow faster and faster.

In conclusion, we can see how the population gains in the rest of Catalonia, those that lead us to talk about the existence of a decentralization process, are due primarily to the output of the metropolitan area population. A population that is living in these territories bordering the metropolitan area but, instead, they continue to have their workplace in the metropolitan area, where they moves each day and where they spend most of their time. Consequently, according to the arguments put forward, we cannot support the idea that there is a decline of the metropolitan area of Barcelona, but rather that city is expanding.

Regarding the main subject of this presentation, the urban mobility and its relationship with the environment, the decentralization process has very important consequences, concerning the increase of the distances people have to travel daily to go working. But also there is another dynamic in the distribution of the population that may have a greater impact on urban mobility patterns: the dispersion, the urban sprawl.

In fact, the process of decentralization that have just seen, has been accompanied at the same time of a dispersal of urbanization, in which the main territories receptors have not been the old industrial tradition cities located in the second ring, but medium-sized and smaller municipalities. At the same time, have become increasingly frequent urban forms based, not in blocks of flats located in cities with a mixture of land uses, but in other types of housing, detached or terraced houses, located in low-density settlements, all of them with residential specialization.

For instance, as shown in Figure 4, while in Barcelona, an example of compact city, only 2,8% of the population live in detached or terraced houses, in the first ring this percentage rises to 12% of population. But above all, we can see how in the second ring, where has arrived much population in recent decades from the core areas, the percentage of population living in single-family typologies is already 38%.

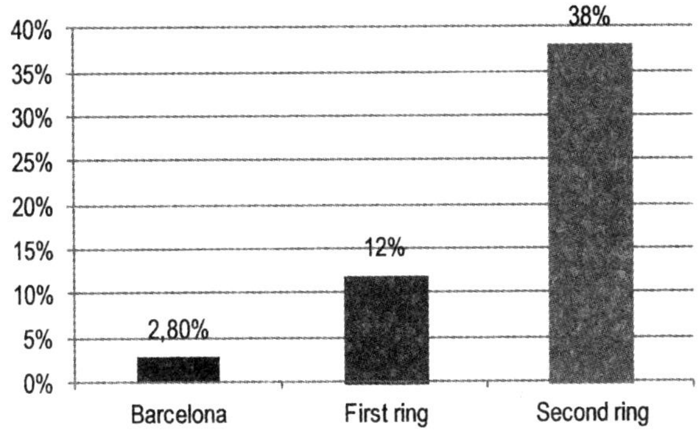

Source: Statistics Institute of Catalonia

Figure 4 Single-family typologies Metropolitan area of Barcelona. 2006

Summarizing, the urban dynamic of decentralization that started taking place in the 1980s in the metropolitan area of Barcelona is combined with a dispersal dynamic from larger cities to midsize and smaller. Moreover, this process has been accompanied by an extraordinary rise in the population living in low-density settlements with a clear residential specialization.

3 URBAN FORM AND URBAN MOBILITY: CAR DEPENDENCE IN THE DISPERSAL METROPOLIS

As is well known, as a result of this urban dynamics urban mobility has been increased, both in terms of the number of displacements and the distance. Furthermore, due to the growth of the population living in low-density dispersed settlements, motorized transport, especially the private car, tend to gain weight in the overall mobility, as it will be described here.

Before starting to analyse the main trends of urban mobility in the metropolitan area of Barcelona it is important to highlight that in this field of study, in the urban mobility, we can find some of the major contradictions of contemporary cities. On the one hand, and from the environment's point of view, urban mobility has clear negative effects, which result in high energy consumption, in air pollution, in emissions at carbon dioxide into the atmosphere, land fragmentation, among the main. On the other hand, the chances of moving in this vast territory are one of the main features of the process of "metropolisation". In fact, urban mobility allows increasing the interrelationships between actors (individuals, companies, and institutions) and better access to services and facilities for the whole population.

A clear example of this duality between the positive and negative effects of urban mobility can be seen in the labour self-containment. The labour self-containment refers the percentage of the employed population of a municipality who works in the same municipality. As shown in Figure 5, this indicator has decreased in recent times, which shows an increasing integration of metropolitan space. In 1995, six of ten employed people had the place of work in their own towns; in the year 2006 are less than half.

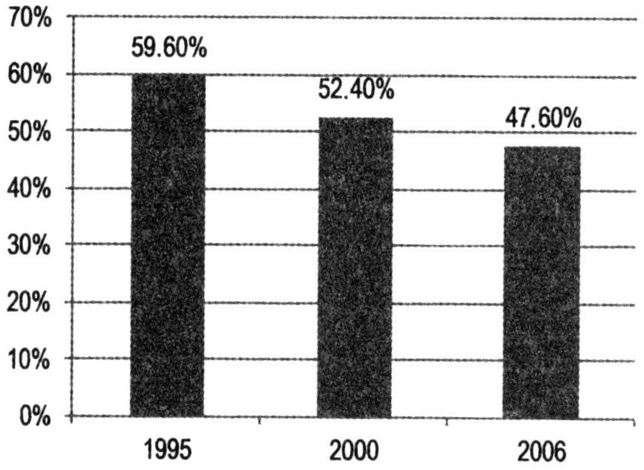

Source: Institut d'Estudis Regionals i Metropolitans de Barcelona.

Figure 5 Labor self-containment of the municipalities. Metropolitan area of Barcelona. 1995 ~ 2006

This greater metropolitan integration has its compensation in an increase in average commuting distances and in a reversal of the possibilities of using non-motorized transport. This increase in the distances has not always translated to an increase in travel time. Indeed, looking at the Figure 6, it can be seen how in 1990 the average commuting time in the metropolitan area of Barcelona was 22.5 minutes, but in 1995 fell to 21.4 minutes. This drop in commuting time was due to the improvement of infrastructure that in the present case were carried out on the occasion of the Olympic Games in Barcelona in 1992.

However, this infrastructure improvement has been offset by strong growth in the number of commuters that have saturated the road network. Moreover, as we have just seen, these commuters are increasingly away from their workplaces. Thus, despite continuing improvements in infrastructure, travel time has been increasing since 2000. Specifically, the average of 21.4 minutes that commuters needed in 1995, have grown to 21.9 minutes in 2000 and to 24.6 minutes in 2006.

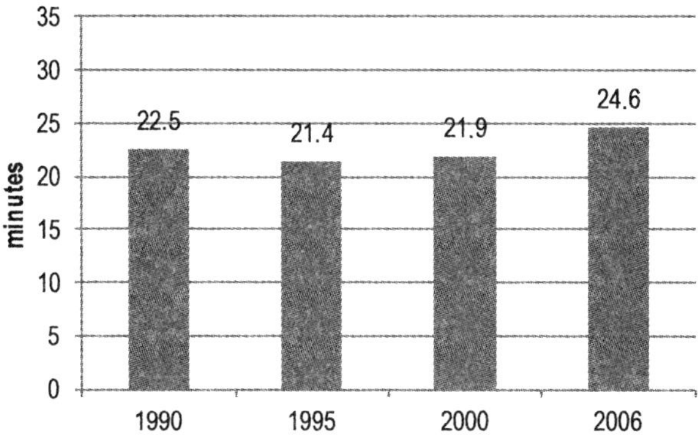

Source: Institut d'Estudis Regionals i Metropolitans de Barcelona.

Figure 6 Average commuting time. Metropolitan area of Barcelona. 1990 ~ 2006

In order to better understand the impacts of urban mobility and its relationship to urban form, we have to study the modal share. Figure 7 shows the commutingmodal share. As shown on the right group of bars, considering all the metropolitan population, the majority uses private motor vehicles, more than half of the population, concretely a 53 %. However, public transport is used by a 26 % , while non-motorized, walking or biking, are used by 18 % of the population.

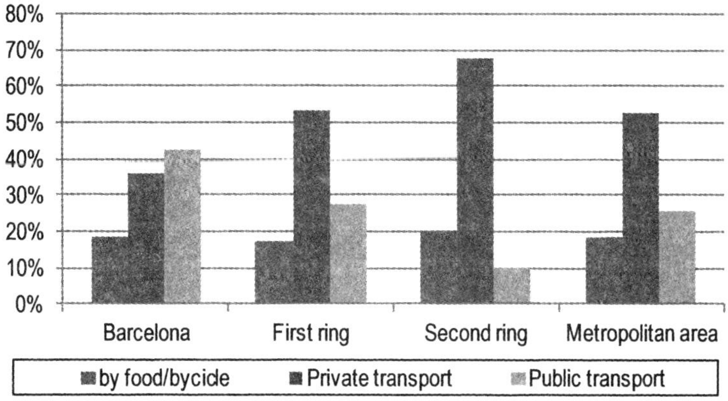

Source: Institut d'Estudis Regionals i Metropolitans de Barcelona.

Figure 7 Modal share. Metropolitan area of Barcelona. 2006

If we analyse the territorial detailed data, we can observe that in the city of Barcelona, an example of reasonably compact city with a dense public transport network, users of this mode of transport increased to 42% , while the private transport fall until the 36%. In contrast, in the first ring and especially in the second ring, where, as already seen, have gained weight forms of low-density development, public transport percentage decreases sharply (27% in the first ring and 10% in the second), while the private sector increases to a remarkable 53% in the first crown and a 67% in the second ring. Consequently, we can conclude that some trends of urban mobility, as for example the distance of displacement and the modal share, are directly related to urban form. In this sense, in the dispersal metropolis both the distance of displacement is longer and the car dependence is higher than in the reasonably compact city.

4 CONCLUSION

It has been seen in this contribution how the process of urban change in the metropolitan area of Barcelona from de 1980s, from an intense concentration of population to an ever faster decentralization, is a sign of the extent of the metropolis and not a symptom of decay. It has also been seen how the process of metropolitan expansion has been characterized by two dynamics: one of decentralization and the other of dispersion, in which especially medium and smaller cities have gained more population. Moreover, in this process, low-density settlements with very sharp residential specializations have gained significant weight.

According with these processes, urban mobility has increased, both as regards the number of shifts and as regards the distance. There has been also an increase of motorized transport, particularly private. From the environment's point of view, as it is well known, this evolution of urban mobility has had some clear negative effects, which result in high energy consumption, in urban pollution, emission of carbon dioxide to the atmosphere, among the major consequences. However, in the urban mobility we can also find positive elements. Thus, increasing the chances of displacement has allowed, among other things, better access to services and facilities for the whole population, and greater interaction between people, companies and institutions.

Thus, in urban mobility we can find two opposing elements: those of negative consequences regarding its impact on the environment, and those that we can consider positive which are at the basis of the existence of the metropolis: the territorial integration. Consequently, if we think about the future of cities and we operate thinking about a new equilibrium between city and environment, one of the main strategies should be to minimize the negative effects of urban mobility but still take advantage of those positive. In this sense, reasonably compact cities with a mixture of land uses, with densities appropriate for the provision of basic services of proximity and well connected by public transport to allow greater access to the entire population, are, without doubt, the best way to promote sustainable urban mobility, with lower environmental impacts.

REFERENCES

[1]BUSQUETS J. Barcelona: La construcción urbanística de una ciudad compacta[M]. Barcelona: Ediciones del Serbal, 2004.

[2]CABRÉ A, MÓDENES J A. Dinàmiques demogràfiques recents a la regió metropolitana de Barcelona[J]. Revista Económica de Catalunya, 1997, 33: 66-76.

[3]CARRERAS J M, OTERO M, RUIZ E. Metropolitan aglomerations[J]. Regió Metropolitana de Barcelona, 2009, 50: 225-246.

[4]CHESHIRE P. A new phase of urban development in western Europe: the evidence from de 1980s [J]. Urban Studies, 1995, 32(7): 1045-1063.

[5]CHESHIRE P, HAY D G. Urban problems in western Europe: an economic analysis[M]. London: Unwin Hyman, 1989.

[6]HALL P G, HAY D. Growth centres in the European urban system[M]. London: Heinemann, 1980.

[7]NELLO O. Urban dynamics, public policies and governance in the Barcelona metropolitan region [C]//MARSHAL T. Transforming Barcelona: the renewal of a European metropolis. London: Routledge, 2004: 27-45.

[8]NELLO O. Spain: Changing century, changing cycle. large Spanish cities on the threshold of the twenty-first century[C]//VAN DEN BERG L, BRAUN E, VAN DER MEER J. National policy responses to urban challenges in Europ. Aldershot: Ashgate, 2007: 333-359.

[9]NELLO O, Donat C. Los efectos territoriales de la crisis económica en la región metropolitana de Barcelona[C]//ALBERTOS J M. Geografia de la crisis económica en España. Valencia: Publications Universistat de València, 2014.

[10]SERRA J. Migracions metropolitanes i desconcentració demogràfica[J]. Revista Económica de Catalunya, 1997, 33: 77-88.

[11]SERRA J. El Territori metropolità de Barcelona. dades bàsiques, evolució recent i perspectives [M]. Barcelona: Mancomunitat de Municipis de l'Àrea Metropolitana de Barcelona, 2003.

[12]UN-HABITAT. State of the world's cities 2008/2009, harmonious cities[S]. London: United Nations Human Settlements, 2008.

[13]UN-HABITAT. Mobility and urban form in planning and design for sustainaible urban mobility: global reports on urban settlements 2013 [S]. London: United Nations Human Settlements, 2013.

[14]VAN DEN BERG L. Urban Europe: a study of growth and decline[M]. Oxford: Pergamon Press, 1982.

Developing a Green-TOD Urban Design Model to Promote Eco-City Development

Kangli Wu

Professor, Harbin Institute of Technology, Department of Architecture, Harbin, Heilongjiang, China; Executive Director, HIT-UC Berkeley/IURD Joint Research Center of Sustainable Urban Development.

Klwu3890302@ gmail. com

ABSTRACT: Promoting Transit-Oriented Development (TOD) has become an important policy goal in many countries. TOD attempts to develop pedestrian friendly environment and attracting compatible mixed-used development around transit station by integrating land use planning and urban design. The concept of TOD is closely related to the concepts of ecological communities and eco-cities strongly promoted by many governmental agencies. Implementing the concept of TOD, therefore, provides a good opportunity to integrate these related concepts for promoting sustainable urban development. However, how these concepts should be coordinated and integrated to develop a Green TOD urban design model that links TOD to ecological community concept still remains a critical research issue demanding in-depth investigation. In view of this, this study attempts to explore the key concepts of Green TOD as well as the urban design approach to implement the concept. The transportation corridor of the underground development of Kaohsiung Rail System is selected as an empirical case setting for its great potential of implementing Green TOD. Employing research methods involving interviews, Fuzzy Delphi survey, and GIS mapping, this paper proposes a Green TOD urban design framework and related planning and design approaches in hope of promoting sustainable urban development.

Keywords: Transit-Oriented Development (TOD), ecological communities, Green TOD, urban design, urban planning

1 INTRODUCTION

Transit-Oriented Development (TOD) has received worldwide attention as a solution of many urban problems such as environmental degradation, automotive abuse, urban sprawl, leapfrog land development in the past two decades. TOD attempts to structure urban spatial development with transit systems and centres middle to high density urban development around rail stations. It also aims at attracting mixed-uses anddiverse land development in and around rail station areas by integrating related land use planning and urban design. By improving the services and functions of transit station areas through related TOD measures, it is anticipated that transit ridership may be increased, and a compact urban form which mutually benefits the city as well as the associated transit system will thus be made possible.

According to previous studies and practical planning experience, TOD should be promoted by coordinating planning measures and strategies in various scales, including regional, metropolitan, corridor, and local (or district) scale. On the local or district scale, TOD aims at creating compact and compatible mixed land use patterns within a walking distance of rail stations. Various daily services, pedestrian friendly environment, and diverse housing and open space are suggested to be provided according to appropriate urban design principles in hope of attracting ample and diversified population, local employment, and urban activities (Cervero, 1998; Wu, 2009).

The concept of TOD, to certain degree, is related to the concept of eco-cities and ecological communities. Promoting the concept of TOD, therefore, provides a good opportunity to integrate relevant ecological community concept which explicitly responses to the sustainable city policy promoted by many government agencies. In view of this, this paper seizes the opportunity of conducting underground development of the rail system and related urban regeneration in Kaohsiung city to propose a Green TOD urban design model as well as to explore the strategies for implementation.

This study uses the Kaohsiung city as an empirical case. Previously a port-based industrial city, Kaohsiung city has now been transformed into a multi-functional metropolis. During the process of adjustment of its urban form, land use patterns and industrial structure, the underground development of the existing rail system in Kaohsiung plays an important role in connecting and sewing urban form and land uses which were previously separated and segregated by the rail system. To foster the sustainable urban development of Kaohsiung, how to develop an operational Green TOD urban design model to help the transformation of the city has also become an important and urgent issue. Given this unique opportunity and urgent need, this paper explores the possibility of developing a new urban design model by integrating TOD concept with the concept of ecological communities. Using the rail corridor which currently conducts underground development as a practical case setting, this paper attempts to address the following two research questions: (1) How should we identify a new framework of urban design based on the Green TOD concept? (2) What key urban design elements and measures should be considered to help implement the concept of Green TOD? Through our empirical research investigation and planning suggestions, it is hoped that the visions of Green TOD and a sustainable city can be gradually obtained.

2 REVIEW OF LITERATURE AND THEORIES

2.1 THEORIES ANDPLANNING PRINCIPLES OF TRANSIT-ORIENTED DEVELOPMENT

Transit-Oriented Development (TOD) attempts to use transit systems (e. g. , rail system or BRT system) as the framework of urban spatial development or redevelopment. It also aims at creating a compact and compatible mixed land development pattern within the reasonable walking distance of a transit station (US DOT, 1989; Calthorpe, 1993; Bernick and Cervero, 1996; Cervero, 1998; Barton, 2000; Wu, 2009). In order to support the operation of a transit system as well as the stations and surrounding urban facilities, adequate urban design is necessary to provide required services and land use patterns that match required living functions and the need of urban development or redevelopment (Bernick and Cervero, 1996; Barton, 2000; Wu, 2009). In addition, the concept of place making and ecological community design should also be considered in order to develop the station areas into a cosy and livable place that is full of cultural characteristics and amenities (Wu, 2009).

Given the interdisciplinary characteristics of the theory of TOD, previous research efforts have addressed several aspects of TOD based on the different viewpoints, such as: (1) researches which explore the relationship between TOD and urban development or urban form (see e. g. , Bernick and Cervero, 1996; Cervero, 1998; Cervero and Day, 2008; Fouracre and Dunkerley, 2003); (2) researches and professional reports which focus on the planning strategies, land use planning, and/or technologies of promoting TOD or transit villages (see e. g. , Cervero, 1998; Wu, 2009; Lee, Yi, and Hong 2013); (3) the economic and financial aspect of promoting TOD or TOD and real estate development (see e. g. , Vickerman, 2008; Cervero and Murakami, 2008); (4) the issues of pedestrian friendly environment design and urban design of TOD (e. g. , Cervero and Kockelman, 1997; Southworth, 2005; Munoz-Raskin, 2009; Wu, 2009).

Inpractice, TOD could be implemented on various scales along with comprehensive and integrated measures from environmental, economic, and planning/design aspects (Bernick and Cervero, 1996; Wu, 2009). On the local scale, it is advocated that business, residence, employment, as well as parks and public facilities should be carefully planned and allocated in the transit servicing corridors by considering the connectivity of trips and the compatibility of activities. Also, the facilities and services of daily life should be accessible within a reasonable walking distance of a transit service.

2.2 FROM 3D TO 6D

According toCervro and Kockelman, a successful TOD must have three key elements (see Figure 1): Density, Diversity and Good Design (Cervero and Kockelman, 1997). "Density" means sufficient land use density and compactness in the transit station areas in order to generate sufficient ridership for a transit system. "Diversity" refers to mixed land use, diverse housing supply, various building, and sufficient public open space and amenities on the neighbourhoods serving by TOD stations. Finally, "Design" means good urban and community design which considers the quality of living environment, pedestrian environment, and place making. The 3D TOD model has been used by a number of studies in promoting TOD and sustainable urban development. On the basis of the 3D TOD concept, Cervero and Murakami (2008) suggested that the 3D model could be extended to 5D in their research conducted by the Institute of Urban and Regional Development (IURD) of UC Berkeley. The added 2Ds are "Distance" (proximity) to a transit system and "Destination Accessibility" respectively. Considering the environment characters and the worldwide trend of urban design, this author proposed the concept of 6D (Figure 1), adding another D: "Distinction" in order to emphasize the importance of distinctiveness and local identity in urban design of TOD station areas (Wu, 2009).

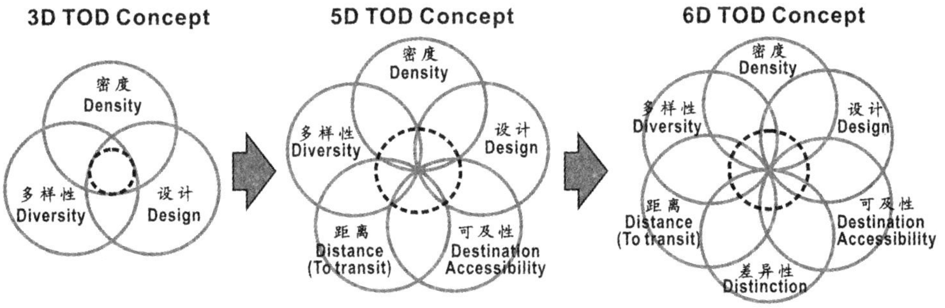

Figure 1 The elements of 3D TOD to 6D TOD(Source: Wu, 2009)

2.3 INTEGRATING TOD WITH THE CONCEPT OF ECOLOGICAL COMMUNITY

This study intends to develop a new urban design approach that integrates the concept of

TOD with ecological communities. The new approach is called Green TOD in the following discussion.

The concept and related research of TOD is reviewed in previous paragraphs, and the concepts of ecological communities are discussed here. In summary, the fundamental ideas of ecological community is to integrate several related concepts, including sustainable urban and local development, ecological planning and design, green transportation, mixed land use planning, diverse and affordable housing supply, health building and health community, short-distance agriculture, environmental justice, public involvement, community management, and community empowerment (see e. g., Calthorpe, 1993; Katz, 1994; Barton, 2000; Beatley, 1998; Roseland, 1998; ODPM, 2004; Wu, 2009). By such multi-dimensional endeavours, ecological community aims to establish a livable and healthy community which helps providing healthy ecological system, daily conveniences, high security, transportation accessibility and sustainable mobility, properly allocation of resources, equity and justice, as well as greater efficiency in administration.

Such a community should also sufficiently utilize regional and local environmental resources and micro-climate characteristics to conduct community planning and design that fit into local environment and culture. Since the above-mentioned concepts, to a certain degree, are related to the concept of TOD, conducting comprehensive planning that integrates these related concepts, therefore, provides a good opportunity to benefit both TOD and associated local communities.

3 RESEARCH DESIGN

This study uses the corridor of the underground development of the rail system of Kaohsiung as the empirical case. This corridor is a pilot region for implementing the concept of TOD and ecological community, thus it provides a good opportunity to address the issues and urban design elements ofpromoting Green TOD (Figure 2).

This research attempts to incorporate both qualitative and quantitative research methods. The research methods employed include field survey, interviews, GIS analysis, and fuzzy questionnaire survey. Interviews were conducted together with a questionnaire study.

The questions addressed in the survey included both closed and open questions in the following two categories:

(1) What urban design framework is required to help promote the concept of Green TOD, and what is the relative importance of the key elements of the framework?

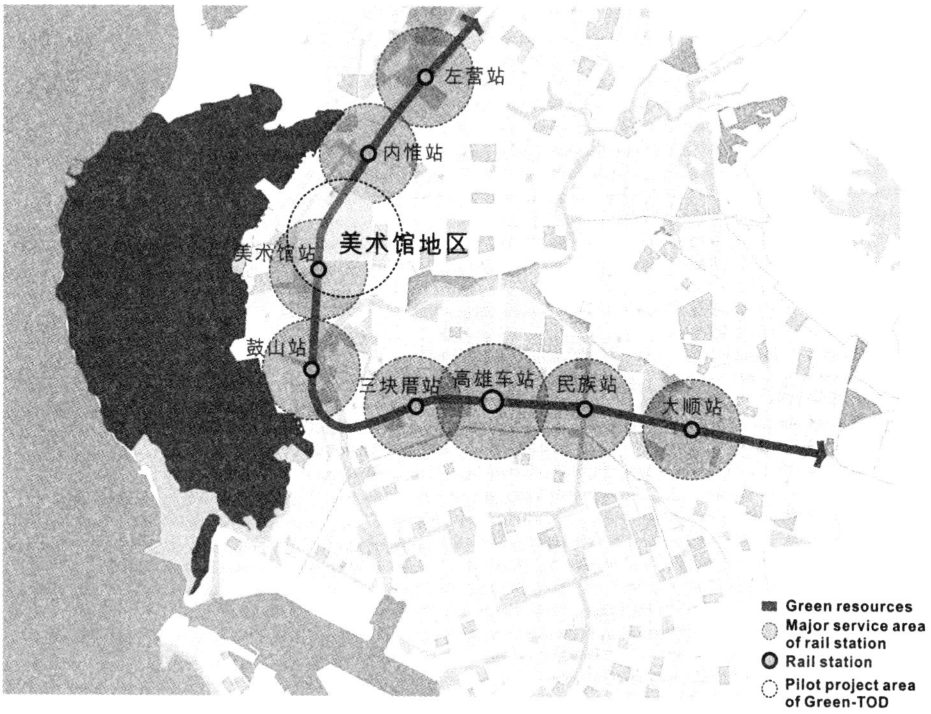

Figure 2 The study areas

(2) What urban design measures and planning strategies are necessary to help implement the concept of Green TOD?

In order to systematically address the proposed research questions, this study uses an interactive planning process, including the following major steps:

(1) Identifying project visions and goals.

(2) Conducting field survey and field observations of the research region.

(3) Developing spatial database for conducting GIS analysis and spatial mapping.

(4) Conducting interviews with the major stakeholders and decision-makers to help identify the urban design framework.

(5) Conducting surveys to test whether the proposed urban design framework and key measures are appropriate.

(6) Developing planning and urban design options and strategies for addressing the key elements of Green TOD.

(7) Adjusting the strategies and urban design content based on the feedback of stakeholders and local people.

4 RESEARCH RESULTS

As mentioned earlier, Green TOD should be implemented by coordinating the measures and planning tools in various scales, including regional and metropolitan scale, corridor scale, district scale, and site scale. This study emphasizes the analysis and planning on the district scale and corridor scale. The following are summaries of our empirical findings.

4.1 DEVELOPING AN URBAN DESIGN FRAMEWORK FOR PROMOTING GREEN TOD

Green TOD requires systematic and comprehensive thinking that allows for investigating the key elements of urban design and addressing the relations among these designelements. Based on the results of literature review and the results of using a Fuzzy Delphi questionnaire survey as well as interviews to explore the opinions of 18 major stakeholders, we propose the following urban design framework for promoting green TOD on the corridor and district level (Figure 3). The tri-angle membership function was used in the fuzzy analysis, and the center of gravity method was used to calculate defuzzy numbers. A α-cut threshold of 0.65 was used to select the items in the proposed urban design framework. The value of defuzzy number is between 0 and 1. The more it approaches to 1, means the more important the item is.

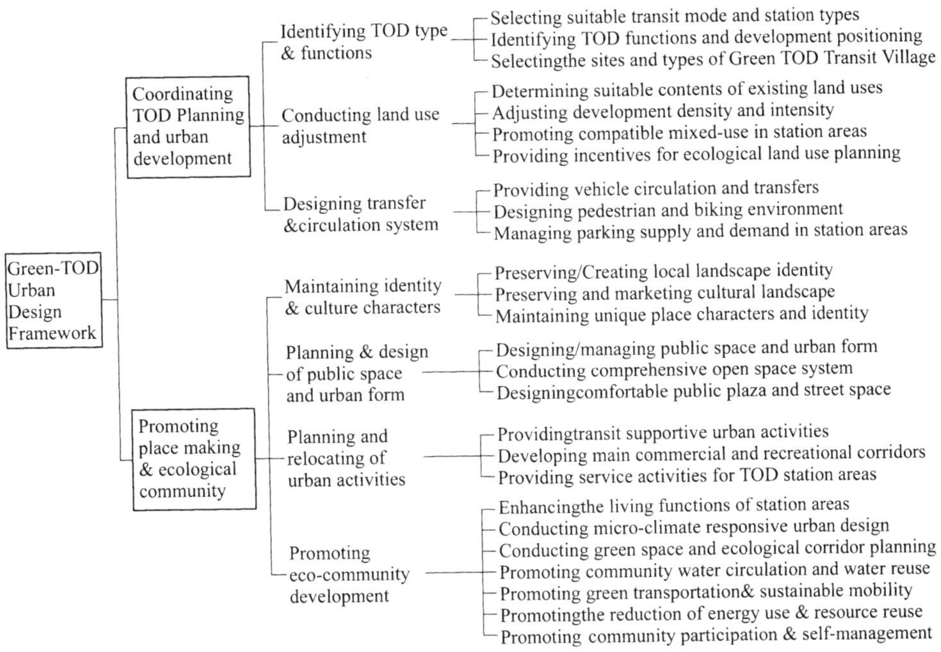

Figure 3 The proposed Green TOD urban design framework

4.2 IDENTIFYING THEMAJOR TYPES AND FUNCTIONS OF GREENTOD TRANSIT VILLAGES

Promoting Green TOD requires a systematic and comprehensive analysis which identifies the basic types and functions of TODplanning units (e.g., transit villages). Based on our field survey, GIS analysis, and interviews with planning experts and related stakeholders, several major types of Green TOD Transit Villages are suggested (Figure 4).

The result also reveals that the multiple environmental characters of the corridor under study which requires suitable planning strategies and action plans that fit into local environmental characteristics and urban context for the selected TOD sites. For example, the newly developed Green TOD transit villages (Type B) should employ technologies of sustainable urban planning and design, and introduce desired service functions and urban activities to serve as a catalyst to foster a compact urban development pattern. As for the Green TOD transit villages in historical inner city areas, the measures of urban redevelopment and regeneration as well as green building technologies should be considered which maintain the unique urban fabric and cultural landscape with collective memory on the one hand, and promote living quality and urban functions on the other.

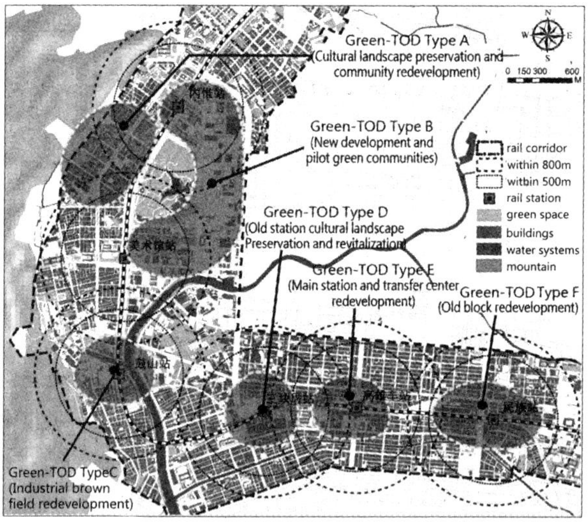

Figure 4 Major types of Green TOD transit villages

4.3 IDENTIFYING THEKEY URBAN DESIGN AND LAND USE PLANNING MEASURES FOR PROMOTING GREENTOD

After exploring the urban design framework and the basic types of Green TOD transit villages, this study explores the opinions of two major decision-making groups in order to help identify the feasible urban design and land use planning measures for implementing Green TOD. The first group studied is the planning professional group, including decision makers, urban planners, and urban designers in related public sectors or planning agencies. The second group includes the selected land developers and real estate developers. A Fuzzy Delphi questionnaire survey was used, and the total effective sample size of the two groups is 34. The research subjects' opinions about the key urban design and land use planning measures are shown in Table 1.

The resultis consistent with some current literature, and shows that the vision of TOD development, mixed land use, adjustment of land use density and intensity, the living function of the station areas, good modal transfer, good open space design, pedestrian friendly environment, and improving the connection to green space system are important urban design elements in promoting Green TOD.

Table 1 The major land use planning and urban design measures

Major Dimensions	Selected measures of urban design and land use planning for promoting Green TOD	Planning Professionals	Developers
Station area and rail corridor development (or redevelopment)	Identifying the visions & positioning of transit villages	0.80	0.77
	Providing incentives for compatible mixed land use	0.74	0.78
	Providing incentives for density & intensity adjustment	0.75	0.81
	Improving the living function of TOD station areas	0.78	0.79
	Providing incentives of promoting ecological land use	0.72	0.75
	Conduciting ecological corridor planning and design	0.74	0.72
	Improving the functions of modal transfer	0.78	0.76
	Improving local open space design	0.76	0.78
Ecological community design and management	Improving pedestrian friendly environment	0.75	0.76
	Enhancing the connection of green space system	0.76	0.78
	Promoting micro-climate responsive urban design	0.73	0.71
	Conducting rain water collection and reuse	0.71	0.70
	Promoting the use of renewable energy	0.72	0.73
	Preserving cultural landscape elements	0.75	0.72
	Promoting vacant(or under-used) land redevelopment	0.74	0.77

Note: the value of defuzzy number is from 0 to 1. The more the value is approaching to 1 means the more important the item is as evaluated by the surveyed research subjects.

4.4 SUMMARIES OF SELECTED URBAN DESIGN MEASURES AND STRATEGIES

After exploring the urban design framework and the key urban design measures, an analysis of selected measures and strategies are summarized as follows:

4.4.1 CONDUCTING LAND USE ADJUSTMENT

According totraditional TOD theory, in order to enhance the node effects of rail stations, the land use density and intensity should be higher in the areas surrounding rail stations (Figure 5), and the land use density and intensity should be decreased as the distance to the rail stations increased (Figure 6). To explore whether the land use condition and development intensity in the rail corridor of the Kaohsiung city is consistent with what is suggested in the traditional TOD theory, the land use intensity of the rail corridor servicing areas was analysed using GIS 3D mapping.

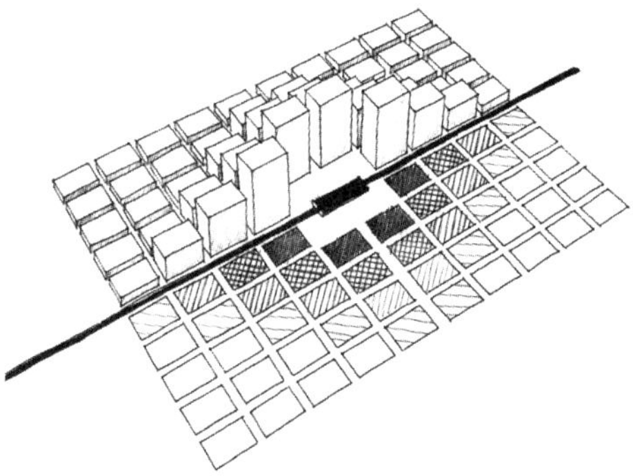

Figure 5 Land-use density and intensity around rail station

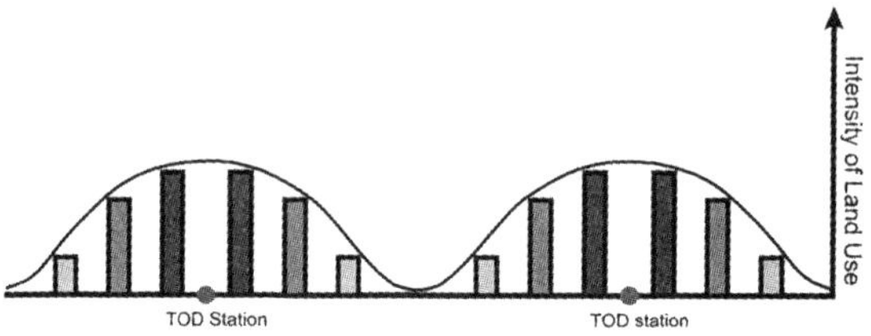

Figure 6 Section of land-use intensity in rail servicing areas

Figure 7 shows that the development intensity (FAR) in the areas serving by the Kaohsiung rail system. The result shows that although the TOD theory suggests that the density and development intensity should reach a peak around the main rail station, the actual FAR distribution in Kaohsiung city reveals that the existing FAR gradient seems doesn't quite follow this rule. This means that the main rail station of Kaohsiung rail system hasn't play an important role in shaping urban spatial development pattern. This situation also reveals that the development intensity (FAR) in certain rail station areas should be increased to enhance the role of rail station areas and to support the functions of rail transit service. Equally important is that the FAR and density in some fast-growing urban fringe areas without transit services should be decreased or controlled to avoid further urban sprawl.

Figure 8 shows similar situation. This figure shows that the spatial distribution of the ratio of existing FAR to allowed FAR (the maximum FAR according to current land use regulations). As the figure reveals, rail station areas, such as the Nei-Wei Station area, Art Museum Station area, and Gu-Shan station area all have a low ratio of existing FAR to allowed FAR which means that there are still capacity for land development. Therefore, related development incentives and infrastructure investment should be provided to encourage future development in these targeted areas, while land development control in some fast-growing urban fringe areas should also be considered in the meantime.

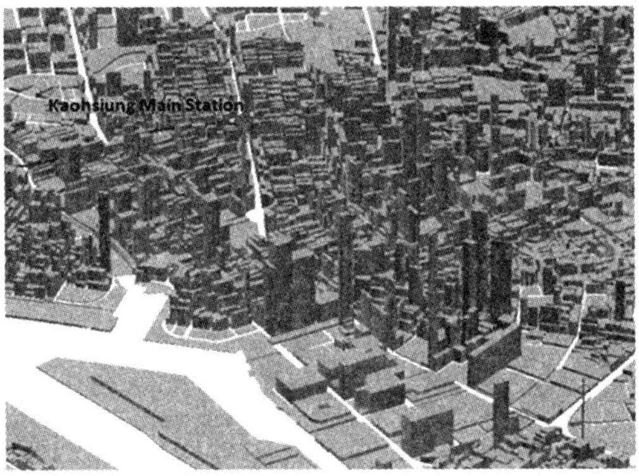

Figure 7 Main rail station and actual FAR in Kaohsiung rail corridor

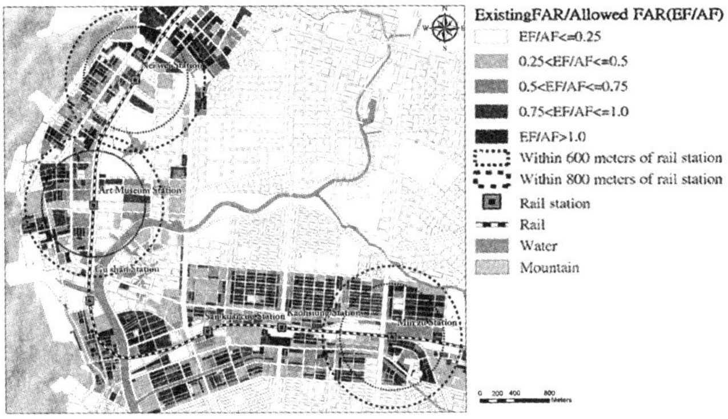

Figure 8 Ratio of allowed FAR to existing FAR around rail stations

In terms of the contents ofcurrent land use and mixed-use situation, the cxisting land use patterns of the nice selected rail station areas are analysed and shown in Figure 9. The results reveal that although mixed land use is common in Kaohsiung city, the content of existing mixed land use of certain rail station areas of Kaohsiung city seems doesn't quite support rail transit services and the diverse living functions as suggested by the proposed Green TOD model. For example, there are still significant amount of land which are used for high pollution industrial use in Gu-Shan station areas (see Figure 9) and there is a lack of public space in many build-up areas around the CBD rail stations. Therefore, promoting more compatible mixed land use and attracted suitable residential and retail activities to certain rail station areas is an urgent task in promoting Green TOD.

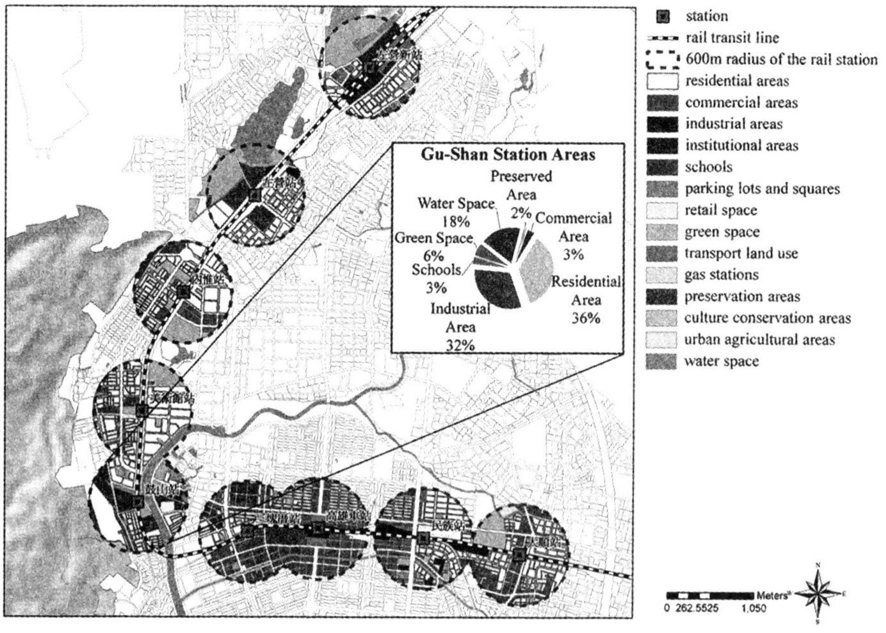

Figure 9 Existing land use patterns in the station areas of rail corridor of Kaohsiung

4.4.2 DEVELOPING MULTIPLE-FUNCTION ECOLOGICAL GREEN CORRIDOR AFTER THE UNDERGROUND DEVELOPMENT OF RAIL SYSTEM

Figure 10 shows the spatial distribution of the green space in the study region. Our analysis shows that the major green space in Kaohsiung City are fragmented and over-concentrated in certain areas. Lack of accessibility and connectivity of green space seems to be a problem in urban green space planning. The underground development of the Kaohsiung rail system provides a unique opportunity to build a green space corridor. With the new green corridor formed after the underground development of existing rail lines (see Figure 11), the urban activities and urban landscape along the rail system could be rearranged to reshape the new functions and images of the city.

To fulfil this goal, some land use control measures (e. g., building setback for construction of green space and pedestrian friendly environment, and providing connected greening courtyard in block design) should be encouraged to connect the new green corridor and the Nei-Wei City Park with the green resources in the surrounding blocks (e. g., see Figure 12 and Figure 13).

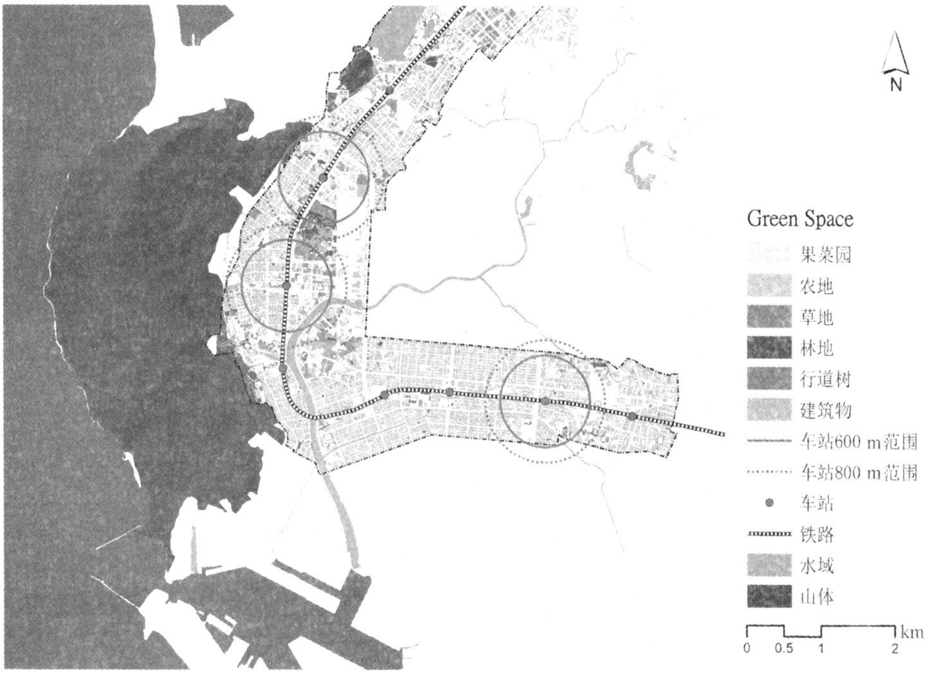

Figure 10 Major green space in the region

Figure 11 Before-and-after analysis of the underground development of the rail corridor

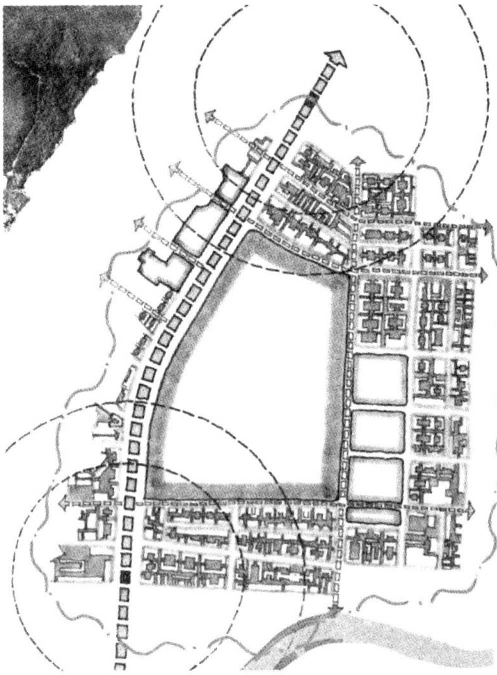

Figure 12　Building green network(Ⅰ)

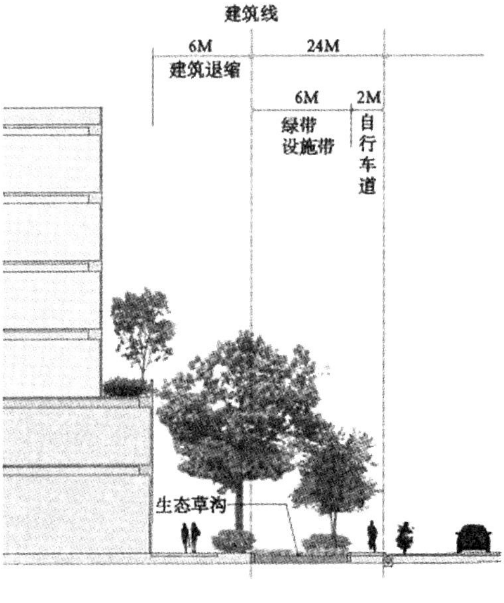

Figure 13　Building green network (Ⅱ)

4.4.3 PROMOTING ECOLOGICAL WATER CIRCULATION AND REUSE

Water collection and reuse is also an important urban design element in Green TOD. Figure 14 and Figure 15 show the concept of regional water circulation and the major sites for urban water retention and rain water collection. It is suggested that water circulation and green networks should be arranged to connect major water collection sites, green space network, and green transport routes (e. g., bike routes and pedestrian networks). The concept of water retention and recycling on the site scale should be extended to block scale or community scale to help implement the concept of ecological water circulation.

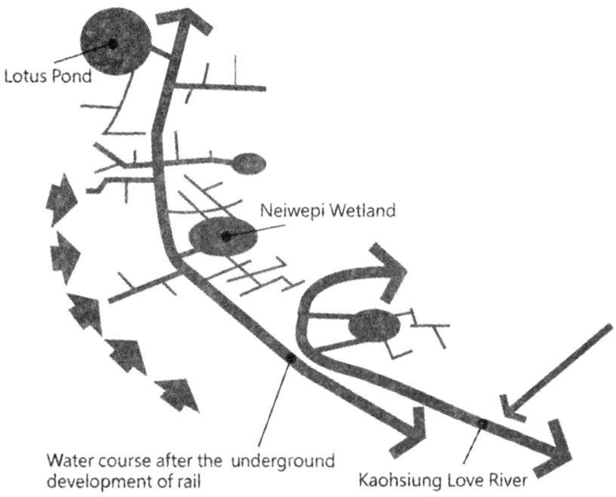

Figure 14 Regional water circulation and rain collection

Figure 15 Establishing green infrastructure network

4.4.4 ANALYSIS OF WIND ENVIRONMENT OF SELECTED STATION AREAS

Green TOD enables the realization of ecological design and climate-sensitive urban design in the station areas. Under such concept, some urban space design and urban design guidelines should be modified according to the analysis of urban heat island effects and micro-climate environment (e. g. wind environment and wind corridor effect). Figure 16 shows part of the results of our analysis of wind environment of certain selected station areas using CFD simulation tools. Based on the empirical findings, some related urban design and land use planning measures are suggested as follows:

Station	Current Condition	CFD Simulation Results
Kaohsiung Museum Station Area		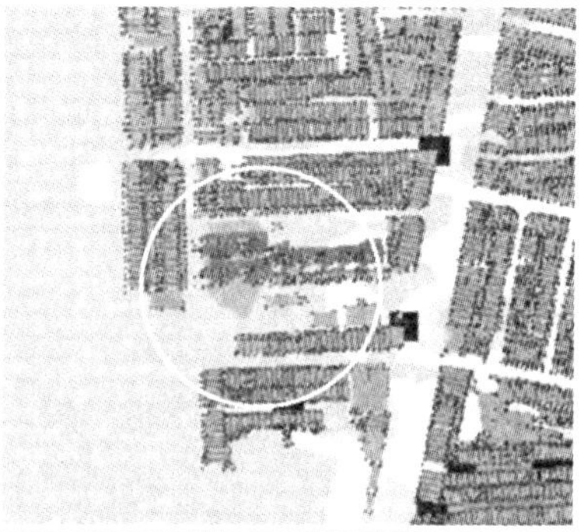
		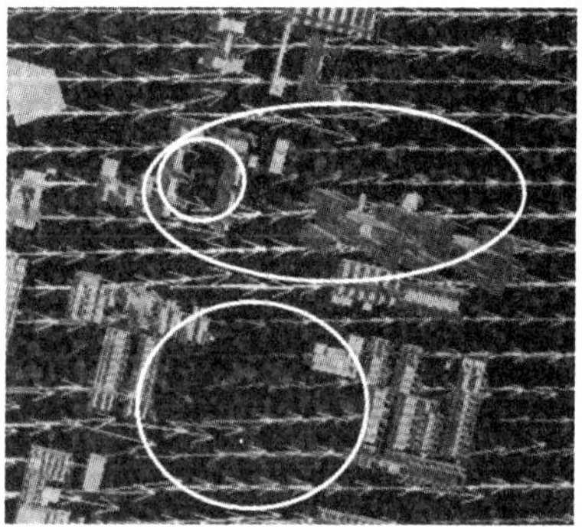

Figure 16 CFD analysis of wind environment

Wind corridor design: Conducting reasonable organization of urban space and building layouts on the block levels, and choosing suitable site plan and street section to control the local wind environment and wind direction that allow comfortable wind passing through communities and building blocks. Key planning considerations include:

(1) Urban green land policy: Better utilizing planting choice and ecological principles to reduce local temperature and heat island effects in summer and weaken the wind force in winter.

(2) Buildingmass control: Control of building form, site plan, building setback, and the organization of building masses to take advantages of the effect of wind channel. And using street planting to create comfort wind environment for residents and pedestrian us-

ers.

In summary, Green TOD should refer to the analysis of micro-climate environment to address the wind corridor effect, building mass control, building setback, urban greening strategies, and design of urban plazas, working together to allow more comfortable and ecological community environment could be obtained. Some design guidelines are suggested as follows (Figure 17):

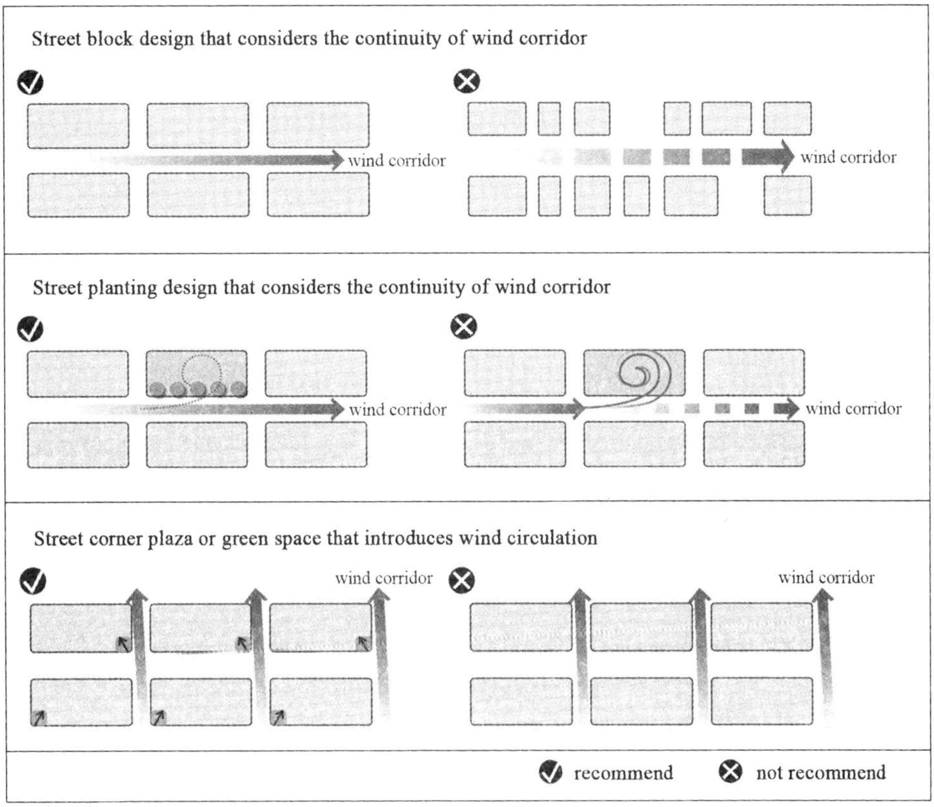

Figure 17 Block level design guidelines based on wind environment

4.4.5 PRESERVATION AND REVITALIZATIONOF HISTORICAL BUILDINGS AND CULTURAL LANDSCAPE

The Green TOD planning in the rail corridor also aims at promoting place making and urban preservation. Place making focuses on creating place identity by preserving cultural landscape elements and by creating comfortable and easy assessable public space. The old station and historical streets with historical meaning and collective memory are preserved together with the efforts of marketing their cultural values. Figure18 shows the concept of the revitalization of the San-Kwai-Tsu old station, and the preservation of some landscape elements, such as old trees, packed park space, and human-scaled small plaza, together with some new design that combines old and new. Creating a meaningful place with landscape and cultural identity will help making the station areas become a popular place for public gathering and communication.

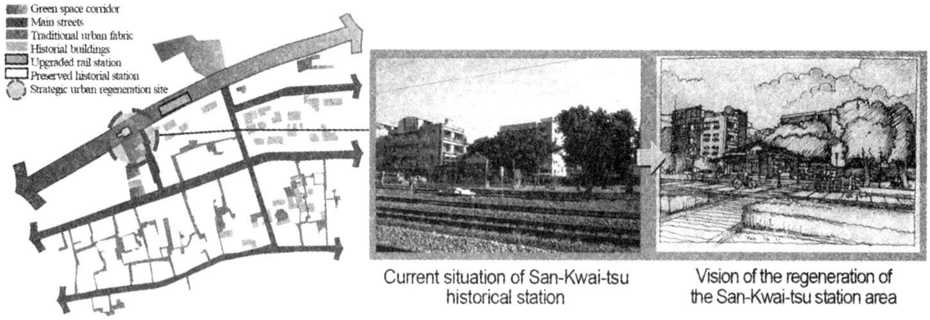

Figure 18 Revitalized historical old stations and historical streets

In addition to preserving some historical rail stations and surrounding open space, the unique urban fabric and the traditional buildings in those station areas are suggested to be preserved (see Figure19). The urban form and the unique architectural style and elements of these historical blocks are suggested to be preserved with suitable building remodelling which maintains the original style and their cultural significance, while the new road construction in city planning should avoid destroying the unique urban fabric and the unique street patterns.

Figure 19 Preserving the urban fabric and historical elements

4.4.6 VISUALIZATION THE URBAN DESIGN VISIONS AND LAND USE ALTERNATIVES

In addition to developing urban design alternatives, visual simulation is also employed in the pre-design stage to help communicate the ideas with the decision-makers and the general public. Some examples are discussed as follows:

1. From back to front after the completion of underground railway.

With the completion of underground railway system and the new green corridor, the building backs on both sides of it will become fronts, which form the connected landscape axes and associated urban activities with new ecological green corridor. The following picture shows the vision of short-term improvement and middle-to- long-term improvement of the building facades along the new green corridor after the underground development of the Kaohsiung railway system (see Figure 20). 3D simulation is used to help decision makers and the general public to visualize the potentials of future development, thereby political and social supports may be enhanced.

Existing situation　　Short-term improvement　　Middle-to-long term development

Figure 20　Simulation of the redevelopment of block facade along the green corridor after the underground development of rail system

2. Simulation of development intensity (FAR).

Summarily, Figure 21 shows the 3D simulation results of various FAR criteria for the corridor between Nei-Wei rail station and Art Museum rail station. This simulation results are shown to the stakeholders (e. g. , decision-makers, developers, and the selected opinion leaders) to explore their opinion regarding what kind of the development intensity and density are accepted by most decision-making groups.

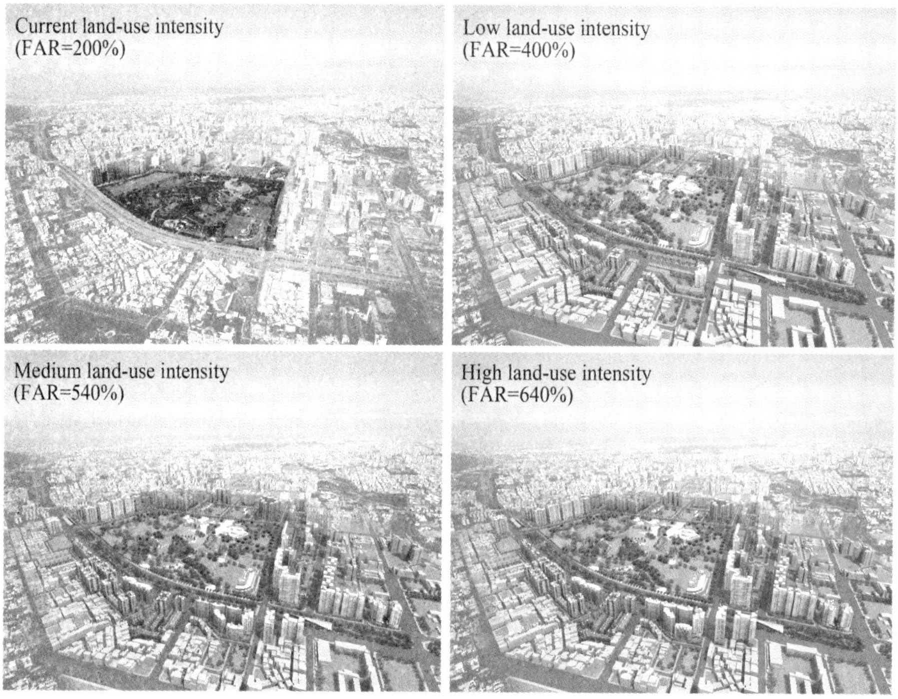

Figure 21 3D simulation of different FAR in the rail corridor between Nei-Wei station and Art Museum station

5 CONCLUSION

This paper proposes an urban design framework and several operational measures for promoting Green TOD in Kaohsiung. The result shows that the city has several potentials in promoting Green TOD. Yet, our research findings also reveal that several current land use situation and urban design contents must be adjusted in order to promote more compact and ecological land use as well as a development pattern that in favour of Green TOD.

In viewing of most current research concerning sustainable and ecological cities tend to concentrate on conceptual planning or policy discussion, which lack of systematic discussion of specific urban design measures and their implementations, this research attempts to establish an operational urban design framework and comprehensive urban design measures to promote Green TOD. The results suggest that a comprehensive railway corridor redevelopment plan should be designated by considering the major key elements of Green TOD, and based on the comprehensive framework and redevelopment plan we could promote strategic planning and urban design actions for developing a Green TOD city.

In terms of the planning and urban design strategies for promoting Green TOD, this paper proposes several strategies, including urban design strategic planning for rail station areas, the concept of multiple-function green corridor design of the underground rail corridor, and readjust of land use planning to promote the concept of ecological communities and ecological cities, urban water retention and recycling, CFD simulation and climate sensitive urban design, urban preservation and revitalization. It is expected that such strategies could reinforce a transformation of the districts along the new green corridor of the railway. In addition, it is also suggested that further research efforts should be made to investigate and evaluate the environment and facility demand of various Green TOD transit villages. This could further facilitate the implementation of the transit villages and ecological communities, which would help shape a sustainable and ecological city.

ACKNOWLEDGEMENTS

This paper is partially funded by a research project of National Science Foundation (NSC99-2410-H-006-081) and a Grant of the Overseas Recruited Scholars Research Funding (AUGA9833216312) of HIT. This author wishes to thank these institutes for their support.

REFERENCES

[1] BARTON H. Sustainable communities: the potential for eco-neighbourhood[M]. London: Earthscan Publications Ltd, 2000.

[2] BEATLEY T. The vision of sustainable communities. in Raymond J. Burby eds. Cooperating with nature: confronting natural hazards with land-use planning for sustainable communities [M]. Washington, D. C.: Joseph Henry Press, 1998.

[3] BERNICK M, CERVERO R. Transit villages in the 21st century[M]. New York: McGraw-Hill Inc., 1996.

[4] CALTHORPE P. The next American metropolis: ecology, community, and the American dream [M]. New York: Princeton Architectural Press, 1993.

[5] CERVERO R, KOCKELMAN K. Travel demand and the 3Ds: density, diversity, and design [J]. Transportation Research, 1997,2(3): 199-219.

[6] CERVERO R. The transit metropolis: a global inquiry[M]. Washington, DC: Island Press, 1998.

[7] CERVERO R, MURAKAMI J. Rail and property development: a model of sustainable transit finance and urbanism[R]. UC BerkeleyCenter for Future Urban Transport: A Volvo Center of Excellence, 2008.

[8] CERVERO R, DAY J. Suburbanization and transit-oriented development in China[J]. Transport Policy, 2008,15(5): 315-323.

[9] FOURACRE P, DUNKERLEY C. Mass rapid transit systems for cities of the developing world [J]. Transport Reviews, 2003,23(3): 299-310.

[10] KATZ P. The new urbanism: toward an architecture of community[M]. New York: McGraw-Hill Inc., 1994.

[11] Lee,Seungil, Changhyo Y, et al. Urban structural hierarchy and the relationship between the ridership of the Seoul Metropolitan Subway and the land-use pattern of the station areas[J]. Cities, 2013,35: 69-37.

[12] MUNOZ-RASKIN R. Walking accessibility to bus rapid transit: does it affect property values? the case of Bogota, Colombia[J]. Transport Policy, 2009,17(2): 72-84.

[13] ODPM (The Office of the Deputy Prime Minister). The Egan review: skill for sustainable communities[M]. London: The Office of the Deputy Prime Minister, 2004.

[14] ROSELAND M. Toward sustainable communities: resources for citizens and their governments [M]. Gabriola Island, BC Canada: New Society Publishers, 1998.

[15] SOURHWORTH M. Designing the walkable city [J]. Journal of Urban Planning and Development, 2005: 246-257.

[16] U. S. Department of Transportation. A guide to land use and public transportation[Z]. U. S. Department of Transportation, 1989.

[17] VICKERMAN R. Transit investment and economic development[J]. Research in Transportation Economics, 2008,23(1): 107-115.

[18] WU K L. The urban design strategic planning of the corridor of the underground development of Kaohsiung railway system[R]. Kaohsiung City Government, 2009.

[19] WU K L. Theory and practice of sustainable and ecological communities[M]. Taipei: Chansbook, 2009.

Urban-Rural Sustainable Development and Planning Method Based on Cybernetics

Huiyi Tang[1,a], *Ming Lu*[2,b]

[1,2]*School of Architecture, Harbin Institute of Technology*

[a]*tanghuiyi@126.com* [b]*hitlm1969@hit.edu.cn*

ABSTRACT: This paper explains the regional/urban-rural planning method based on systematics and cybernetics. In order to deal with the complex problems which are faced by the urban-rural sustainable development, it is urgently necessary to clarify a sort of scientific planning method. This method is based on the Metasynthesis Engineering method of Qian Xuesen School of systems science, adapted to dealing with the "Open Complex Giant Systems" (OCGS).

The urban-rural synthetic development planning is the basis of physical environment planning: its relevant factors, capable to influence the urban-rural system as much as the geographic and social ones, require that the Urban-rural Panning and Design Institute work co-ordinately with the Department of Integrative Design System for Socialist Construction to make dynamic planning and timely supervision and control, in order to guarantee the scientific and democratic decision making mechanism.

Key Words: systematics and cybernetics, open complex giant system, planning method of systems engineering, scientific and democratic decision making mechanism, department of integrative design system

1 INTRODUCTION

The city planning process regulated by the British planning pioneer Patrick Geddes during the early 20th century (1915), based on investigation-analysis-plan scheme, transformed into the British Systemic Planning or Structure Planning implemented in the 60s of 20th century, structuring computer models in concern with certain goals, objectives and projects in order to control the development of the system.

Now we are facing the complex situation of a rapid development, according to the Metasynthesis Engineering methodology of the Qian Xuesen School of Systems Science, which advocate the Systems Engineering or cybernetic planning method in order to deal with the Open Complex Giant Systems (OCGS) in a holistic and systematic way.

2 URBAN-RURAL SUSTAINABLE DEVELOPMENT

China requires sustainable development strategies that conceive urban and rural area as a whole system. China is facing a series of serious problems, as ecological system vulnerability, environmental pollution, resources exhaustion especially water, people migration from rural area to urban one, population and transportation congestion in the cities, as well as various urban diseases. The solution of such problems of course cannot be achieved without the urban-rural planning institute, but it does not suffice.

The nation has to have master plan or top design, and the urban-rural planning and design institutes needs to work in a coordinated manner with it.

3 REGIONAL/URBAN-RURAL SYNTHETIC DEVELOPMENT PLANNING

The urban-rural synthetic development planning is the basis of physical environment planning: it determines the goals and objectives of planning according to the system structure and function. For instance, the synthetic planning of the Central Plains Economic Zone determines the following goals and objectives (subsystems): agriculture modernization, new industrialization mode, new urbanization mode, infrastructure construction, social guaranteeing system construction, culture inheritance and innovation, ecological system improvement, environmental protection, disaster prevention.

The urban-rural synthetic development planning might include the following goals and objectives: developing satellite cities system and anti-magnet centres in order to manage the population pressure through a better balanced distribution in the territory; supporting ancient city and cultural relics protection; redistributing the activities concentrated in the urban centre; implementing industrial transformation and up-grade; developing effective measures to integrate migrated workers; developing public service and welfare policies; protecting and improving the ecological systems, improving the environmental pollution treatment, adopting measures for disasters prevention, agriculture modernization and new countryside construction.

4 SOCIAL SYSTEM-GEOGRAPHIC SYSTEM-URBAN-RURAL SYSTEM

The social system consists of 4 aspects and 10 subsystems:

Physical civilization——economy, people's physical quality;

Spiritual civilization——thoughts, culture;
Political civilization——democracy, legislature, institution;
Geological environment——ecological system and environment, physical environment, infrastructure.

The geographic system, consists of 8 subsystems, i. e. population, resources, industries, infrastructures, cities and towns, ecosystem, environment, disasters prevention.

The urban-rural system includes the 8 geographic subsystems just mentioned, and further breaks into more detailed subsystems as residential areas, public service facilities, culture, transportation, greenery, etc.

5 THE PLANNING METHOD OF SYSTEMS ENGINEERING

It is necessary to base the urban-rural planning on the special character of the existing cities, which originates from their historic evolution and according to its systematic nature, in order to determine goals and objectives.

The "Systems Engineering Planning Method" in dealing with the OCGS has, first of all, to analyse the system's structure, function, developing regulations or laws, dynamical mechanisms, chances and challenges which it is facing, as well as the way of optimization control. The method of analysis is the method of systematics which is the dialectical unification of reductionism and holism.

The analysis has to move from a macroscopic qualitative approach to a microscopic quantitative one, to make the utmost use of web database, taking advantage of the traditional sciences as mathematic as much as possible, and of the innovative technical devices (satellite observations, GIS, etc.).

One of the most important forms of control is to do away with the binding forces which hinder the social development, to design the institution, in order to release social vitality, i. e. reform the relationship between production and superstructure, in order to emancipate the productive force. This is an innovative developing mode where top down and bottom up initiatives work together, to cause the society continually emerging new functions.

To emancipate at the microscopic level while control at the macroscopic one: this is a relationship of dialectical unification. Emancipation also could cause chaos, while necessary macroscopic controlling measure has to be taken, but it must handle the appropriateness well enough, because innovation usually emerge at the brink of chaos.

This notation of systems engineering, surpassed the systems engineering method that overrelies on mathematics and computers, and rather, with human-computer in combination while human plays the main role, this is the Metasynthesis Engineering method, conducted with Marxist Philosophy and needs broad coordination.

This method has to grasp major contradiction and the major aspect of contradiction smartly, focus attention on foreseeing problems, problems which is easily to be neglected, and the potential problems.

For example, the livable city planning has to solve a series of problems as city sustainable development, environment, residential district planning, housing provision, public service, infrastructure, management, etc..

For this end, it is necessary to analyse the relevant factors from the viewpoint of urban-rural system, geological system and social system, in order to determine the goals, objectives and guaranteeing measures of planning, while for the guaranteeing housing construction, it is necessary to especially focus on the quality of slum area renovation plans, and problems related to public services and their management.

6 URBAN-RURAL PLANNING AND DESIGN AND DEPARTMENT OF INTEGRATIVE DESIGN SYSTEM FOR SOCIALIST CONSTRUCTION

The Urban-rural Planning and Design Institute has its own specialty work, namely, the physical environment planning and design or land-use and transportation planning which cannot be substituted by any others. The global problems of economic and social development need to be unified handled, planning and controlled by the Department of Integrative Design System for Socialist Construction, in order to acquire the most optimal development. The urban-rural planning and design institute works in close coordination with it, and at the same time, work in coordination with the urban-rural government, to make dynamical planning with timely supervision, surveying and control, and guarantees the scientific and democratic decision making mechanism of planning.

NOTES

(1) Metasynthesis Engineering method of Qian Xuesen School of Systems Science, was

advocated by China's great scientist Qian Xuesen for dealing with the Open Complex Giant Systems (OCGS) which, as he classified, as the highest complex sort of systems, such as the social system, human brain, etc., in contrast with the simple systems as a simple device, or simple giant systems as the laser system. This method requires the combination of human-computer with human playing the main role, and the full use of human image thought and inspiration thought as well as the logical thought, and the full use of human knowledge.

(2) Department of Integrative Design System for Socialist Construction is the institute system advocated by Qian Xuesen for implementing the social engineering task. Now in China, the Leading Group of Overall Deepening Reformation would include such sort of institute.

(3) New industrialization mode now in China means a new mode of industrialization differentiated from the old mode which is with low efficiency and high energy exhaustion and high pollution. The new mode is intended to be in higher attached values, high efficiency, green production, low carbon economy and cycling economy.

(4) New urbanization mode now in China means a new mode of urbanization differentiated from the old mode which was focused on land development. Instead, the new mode would focus on rationally dealing with the rural migrated population with their employment, living and social guaranteeing problems.

REFERENCES

[1]BAO Shixing, GU Mengchao, TU Yuanji, et al. Macroscopic architecture and microscopic architecture[M]. Hangzhou: Hangzhou Press, 2001.

[2]Shanghai Jiaotong University. The key of wisdom: Qian Xuesen on systems science[M]. Shanghai: Shanghai Jiaotong University Press, 2005.

[3]TANG Huiyi, LU Ming. Science of city planning (broadened): urban-rural planning systems engineering[M]. Harbin: HIT Press, 2013.

[4]TANG Huiyi. Systematology: basic theory for scientific development of social system (to be published).

The Peri-Urban Landscape as Battlefield Between Private and Public Perspectives

Alessandra Lai
Department of Architecture Planning & Design-University. of Sassari (Italy)
laissandra@ gmail. com

ABSTRACT: The last forty years have witnessed expanding urban settlements spread into the countryside in a wide range of forms. These are the expression of an increasing number of autonomous visions projected on the territory and reflect a growing variety of lifestyles and ways to occupy spaces. The dilemmas related to the crisis that affects the relationship between city and countryside have deep roots that need to be understood through a holistic approach. The established "habit" of dealing with all matters linked to the territorial transformations only from an "urban-centred" point of view should be challenged.

The study presented here gives an account of some significant steps that have irreversibly marked the relationship of proximity between city and countryside. These steps are presented from a holistic point of view, linking economic, social and physical transformations.

A brief overview of growth processes in the metropolitan region of Cagliari and the progressive abandonment of agricultural activities in Sassari's olive tree belt is presented.

The following section addresses the measures that have been implemented to reverse the negative trends observed in these peri-urban regions, including a brief assessment of the effects of the policies.

Finally, some conclusions are presented connecting with the topics of fairness and sustainability in regards to land use change and the on-going conflict between two different perspectives or concept ualizations on the future of the environment in the fringes of the two main Sardinian cities. This will suggest that the case is not confined to the traditional opposition between private profit and public interest, but it rather involves public profits as well.

Key Words: peri-urban landscape, Cagliari, Sassari, green belt, private-oriented development, agricultural landscape

1 INTRODUCTION

The harmonious "opposition" between city and countryside, admirably represented in the fresco by Ambrogio Lorenzetti The Effects of Good Government in Town and Countryside (Figure 1), part of the cycle of frescoes depicting The Allegory and Effects of Good and Bad Government (Sala dei Nove the Palazzo Pubblico-Siena 1338-1339) represents the aspiration to an ideal reality, perhaps possible, that only the good government could accomplish. It was conceived as an exhortation to choose the right path that can lead to prosperity and harmony.

Figure 1 The Effects of Good Government in Town and Countryside by Ambrogio Lorenzetti

Like many iconographic representations with the bucolic countryside in the background of the merchant city (see Francesco Rosselli, Nicolas Poussin, Giorgione, Annibale Carracci, Peter Paul Rubens et al.), Lorenzetti's work has contributed to build up the myth of a harmonious coexistence between city and countryside, chronologically placed in an idyllic past that had never existed in reality. Indeed, if we look back into urban history, we realise that the dilemmas that we are facing nowadays are nothing new: the forms of mutual interference of the countryside and the city must have been known since Roman times. For instance, looking at the suburbs of Rome and the endless sprawl of farms, Dionysius of Halicarnassus noted, at the time of Augustus, that so closely connected were the city and the country, that it gave the beholder the impression of a city stretching out indefinitely (Morley 2002, page 33). Nevertheless, in both the Greek and Roman civilizations, the fields were regarded as having equal importance to the streets and houses for satisfying the needs for living. The countryside surrounding the city was consciously considered an extension of that. It was the saltus, the wild and unproductive land, that was to be regarded with disdain.

This perspective changes when we look at Northern Europe: unlike the Mediterranean cultures, the livelihoods of the northern populations were based on wild nature, managed as a common good, and thus not restricted to private property, as it was in Roman settlements. In all cases land was a commodity that everybody valued, whether hunter, nomad or city-dweller. Differences were only in the way each civilization used it (Steel, 2013).

Today, the productive lands surrounding the cities of ancient formation, which have historically grown up from the relationship of interdependence between the fertility of the soil and the presence of water, are invaded by other functions that are not entitled to a better compatibility, but rather are regulated by "advantageous position".

2 WHAT HAD BROKEN THE INTERDEPENDENCE BETWEEN CITY AND COUNTRYSIDE?

Although not comparable with the aesthetic ideal of the Romantic paintings, the relationship between town and countryside in the past has expressed an ecological dimension of living. According to the ecological model in the previous modernity (Mattei, 2012), town and countryside were organized around a community structure in equilibrium, where everything was in balance with its parts and with the available resources. At a certain point, something happened and the balanced relationship between city and countryside went through a crisis that seems endless. Some scholars use to identify the reason of this "break" in the Enclosure Acts, especially in England, (Boyle, 2003; Gonner, 1912). For the Southern territories it is a mix of causes, lead by the industrialization process.

In England, after the Civil War, the enclosure of common lands, which had been practiced in an informal way by that time, was encouraged by the Parliament as an answer to the increasing demand for feeding the growing cities, as the traditional management of land was not able to (Hoskins, 1955). The enlarged size of each plot of land and the investments on the experimentation of new techniques generated an increase of the agricultural labour productivity by 90% between 1700 and 1800. On the other hand, the active population in agriculture went from 70% to 37%. The majority of peasants and small landowners did not have enough resources to realize the fences or to pay the legal fees that the enclosure acts demanded, so the concentration of land ownership went in the hands of the English aristocracy.

Something similar occurred in Sardinia, more than a hundred years later, with the "Edict of Locking System" (L'Editto delle chiudende), which allocated the land to the people who were able to fence it. The abolition of the common uses caused significant inequalities that were not anticipated, and the population began to oppose it with determination, occasionally with violence (Pirastu, 1973). The most popular uprising was known as the "su connottu", the "known", in which the rebels were asking to return "to the known", that is the tradition and hence the common use of the land.

When rural workers, excluded from the management of agricultural land, moved into the city, the social structure that supported rural communities began to crumble. The gap between the feeders and the fed was widening and it was about to get a whole lot wider (Steel, 2013).

But the wave of enclosures marked another critical step, highlighted by Ugo Mattei in his manifesto on the commons. Before the "fences", the three main political subjects had a correspondence with as many proprietary models: the sovereign as the eminent owner of the entire national territory; the lords as private landowners; and the commoners, those people who "owned" common land and therefore only "had" common goods. After the "fences", the models were reduced to only two: the one of the sovereign state and that of private property. The majority lost its one.

The change was radical: it meant passing from one type of existence that was qualitatively collective to another one that was quantitatively individualized. Modern state and private property became natural allies against the commons (Mattei, 2011).

The leap forward caused by innovations in the field of transport made manifest that the physical tear of the link between the city and the surrounding territory was determined, foremost, by the railway. The ability to bring food virtually anywhere broke the chain of dependence from the rural areas surrounding the city, automatically devaluating them. City and countryside seemed not to need each other anymore. The improvements in transport and communication implied the annihilation of space through time.

3 RECENT FORMS OF CONTAMINATION BETWEEN CITY AND COUNTRYSIDE

The innovations in the field of mobility for people and goods have accelerated the on going social revolution, influencing the forms of organization of the living space of the city and the countryside. This, in particular, became exposed with the removal of the protection system derived from the "commons". The country appears as empty: perspectives on his fate are no longer the projection of common visions, but rather of individual ones, its value is no longer measured in terms of quality but quantity.

The city reshapes its morphology and its dynamics: a new middle class emerges energetically and exercises its influence, choosing to build its own city in areas not previously urbanized (Secchi, 2013). At the same time, the rural workers, who move towards the mass production centres, determine a sudden high demand for residential buildings, that will be interpreted according to very different types. The less expensive peripheral areas are designed to absorb the demographic spikes.

The service sector in contemporary history has further eroded the boundaries so strongly stigmatized in paintings until the first half of the Nineteenth Century that saw the walled city in non-confrontational opposition to the natural and cultivated areas. The natural crown that encircled the city is performing the same functions that had justified the cities birth, that is the function of the exchange and trade, resulting in depletion of commercial relations within it (Berdini, 2013). The countryside is no longer the counter-space of the city, meant as the place for production (agricultural or industrial) that justifies and allows its existence, but rather the cross-contamination starts between the two fundamental resources for a living.

The city is branching out: first, it develops according to preferential guidelines, like those provided by transport infrastructure, but still retaining a certain, "relative" building density; more recently it spreads in an apparently random form which, at least in the Italian case, started in the 1960s, years of the so-called "economic miracle".

The phenomenon of urban sprawl, which is easily observed in satellite pictures captured at night-time making Earth look like a starry sky, does not reveal any preferences or rather, does not allow us to recognize common strategies of spatial organization. The forms of dissemination recount individual strategies or, at the most, micro-community ones. It lays bare the concept of undifferentiated space that penetrates the culture of living up to the individual dimension. The abstraction of the homogeneous space, assumed for simplification in some models, becomes, in some processing, abstraction and simplification to be in a state of reality. According to this conception: anything can be located anywhere (Indovina, 2003).

The city in transition from Fordism to the service-based model of the metropolis sees the spatial organization changing according to the transformation of the economic structure.

If industry was the first to characterize the large city, today this activity can also be located elsewhere, outside the boundaries of the city, region, or the state, while the city surroundings is "fertile" ground for the proliferation of tertiary businesses, often small to medium in size. In cities they have been always produced fewer objects and more direct services, as well as to the person, to companies (Chiodini, 2009).

4 WHAT HAPPENS TO LOCAL SCALE: TWO PARADIGMATIC CASES IN SARDINIA

The choice of case studies presented in this chapter responded to two paradigmatic contexts with respect to the relational dynamics between city and countryside, because their stories have reshaped the urban landscapes.

One case represents the emblematic expression of the complicity between state sovereignty and private property-this is the case of the wide area of Cagliari (let's call it metropolitan even if this term is no longer suitable in terms of official nomenclature), in southern Sardinia-the other shows the effects of the private-oriented development-the case of Sassari and the suburban belt of olive trees in northern Sardinia. In both cases the growing pressure of building on agricultural land that made up the green belts of the two cities can not be merely conceived as a response to the shortage of housing supply, because a substantial part of the existing housing stock is vacant. It is clear that the reasons must be sought elsewhere.

4.1 The Southern Sardinia

It might be better to introduce some features of the wider regional context the two cases of study belong to.

Sardinia is the second Italian island in terms of area. Its low density settlement system is made by 377 municipalities, 80% of which has less than 5,000 inhabitants (Figure 2). Currently there are 1.648 million residents to which are added during the summer about 8 million tourists, almost exclusively along the coast, where there are placed the majority of second homes and resorts, as well as the industrial areas of basic. The population density of residents remained almost stable in the last 30 years, around 69 inhabitants/ km^2, varying, according to the old set-up of the provinces, between the 38 of Nuoro and the 110 of Cagliari and confirming Sardinia as the Italian region with the lowest population density (Istat, 2014). Cagliari is the capital of the region and the first in terms of population with its 154,000 inhabitants, which become 420,000 if we look at the wide area that includes other 15 municipalities distributed along two imaginary concentric circle centred in Cagliari. (Figure 3)

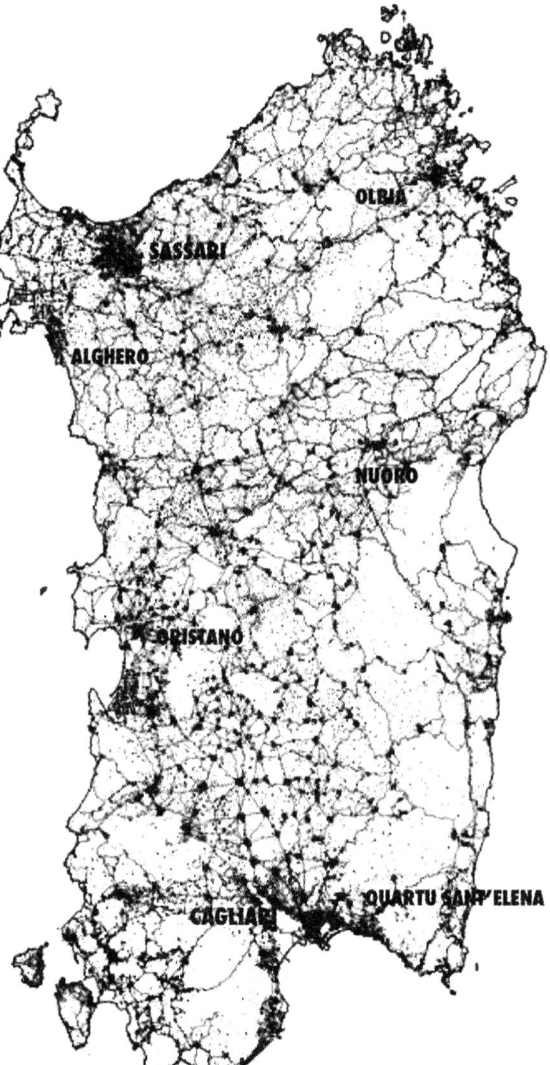

Figure 2 Map of the municipalities and the road network of Sardinia

Sassari is the capital of the same name province and reaches almost 128,000 inhabitants. Since the logistic features are very different, Sassari and Cagliari have been developed following two different models as the following pages show, significantly connoted by their residential area expansion. In both cases the main contradiction lie on fact that the new houses have been built in the recent decades while the population was not increasing. (Figure 4)

Indeed, even if the population of Sardinia has remained more or less stable over the past 20 years (only the immigration phenomenon is giving the island some percentage points of increase), the numbers of building permits for new residences is one of the highest in Italy (Figure 4).

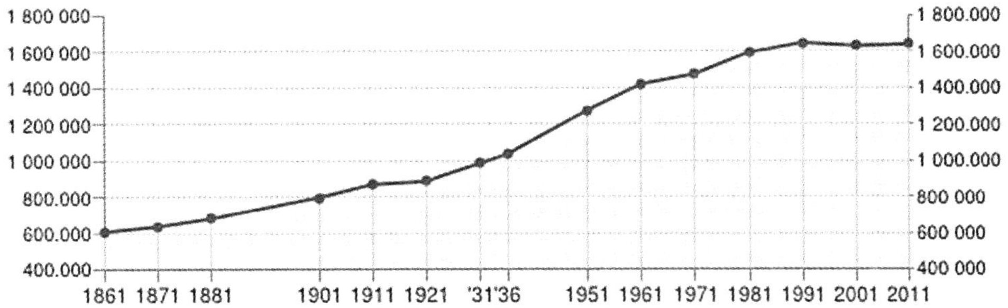

Figure 3 Sardinia Resident Population 1861 ~ 2011 (Istat, 2013)

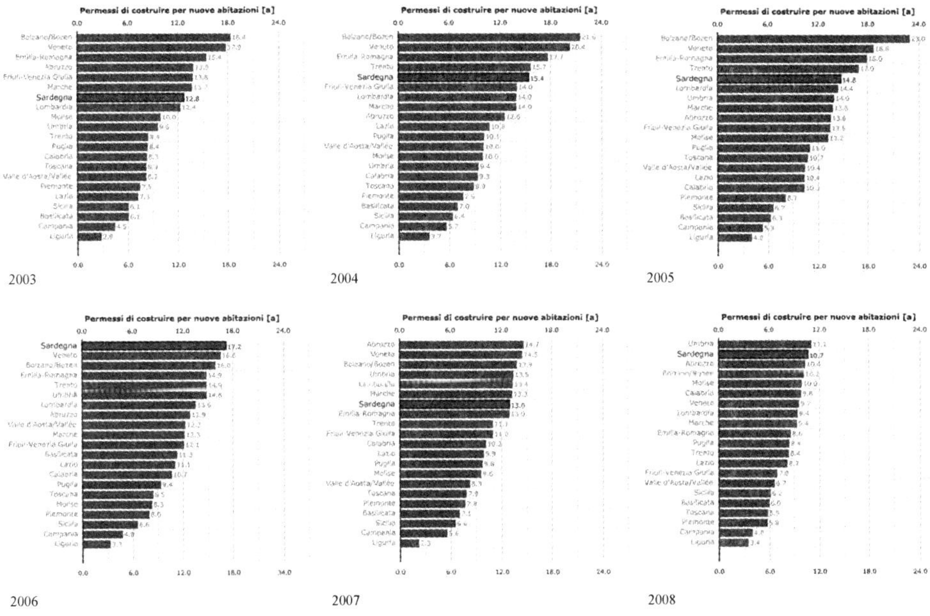

Figure 4 Number of new residential building permits/1000 families (Istat, 2014)

4.2 THE WIDER AREA OF CAGLIARI

Cagliari, the capital of the region, is surrounded by a ring of smaller towns born as an aggregation of independent villas for the management of agricultural land around and behind them: Monserrato, Selargius, Quartucciu and Quartu Sant'Elena. They have grown formally independent, but administratively linked to the capital.

Paradoxically, since they became autonomous municipalities (around 1983 ~ 1990), they have been losing their particular character, the one that allowed their clear identification as towns. Today, were it not for the street panels, it would be difficult to distinguish the end of one and the beginning of the other. The residential and commercial settlements are almost filling in every gap (Figure 5).

Figure 5 Maps of the urban development in the Metropolitan area of Cagliari

Many of the new residential areas, targeted at young families, have occupied the suburban agricultural land which, until recently, had been working as an eco-border (Landscape Regional Plan of Sardinia, 2006).

The image on the left of Figure 6 reveals that "young families" have moved to the borders of the small towns that surround Cagliari, finding their home in the recently erected buildings, increasing the number of movements to the city of Cagliari, which represents the major attractor of traffic for study and work.

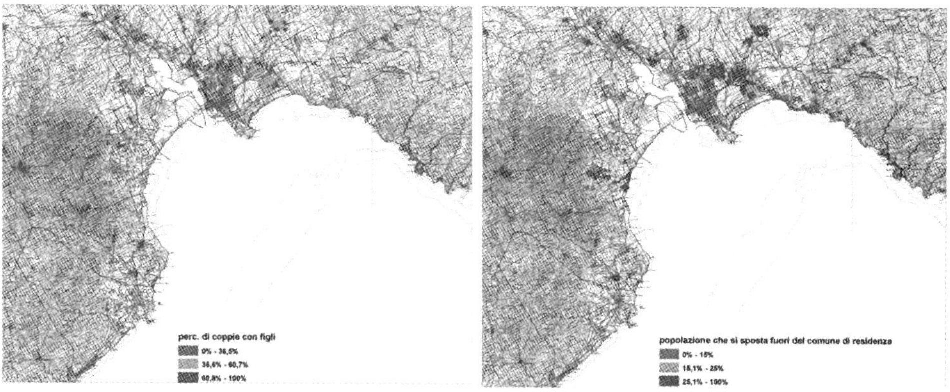

Distribution of couples with children   Population that moves out of the town of residence

Figure 6 Edited by the team "Housing"-Strategical Plan of the Wide Area of Cagliari (2007)

In terms of geographic distribution, the young families seem to be somehow "giving their back" to the town centres, in correspondence with the attitude revealed by the administrators of the small municipalities: "Yes we have attracted some new population but it is not integrated with the local community, they do not take part in the social, cultural and commercial activities, and they use to prefer to enrol their children in the kindergarten next to their workplace".

As in many other parts of Europe, local administrations have promoted the development of residential, retail and industrial areas as a means to increase their cash income through taxes. The shared strategy to attract new population is based on mostly luring young couples, with or without children with:

- low cost for square meter;
- reserved parking close to home;
- presence of terrace or small garden.

Sadly, these are the only factors that "inspire" the new way of building. And that is the reason why new settlements lack architectural quality and care for the components of the urban and natural environment. They are successful because they satisfy the demands for a product that housing in Cagliari or in the historic town centre cannot deliver.

If we approach the analysis of this phenomena focusing on a set of common values, as the efficient and sustainable use of our resources improving the quality of the existing ones, instead of generating new artificial and fragmented ones, or the quality of the air and the urban landscape, we can figure out how those short term strategies affect negatively our living space :

- in spite of a stock of over 5,000 empty homes in the capital, residential developments keep being built;

- an increasing volume of incoming traffic towards the main service centre, Cagliari, is recorded;

- the interstitial voids between the various towns are perceived by Authorities and residents as lacking in quality, just as empty spaces waiting to be filled. (Figure 7)

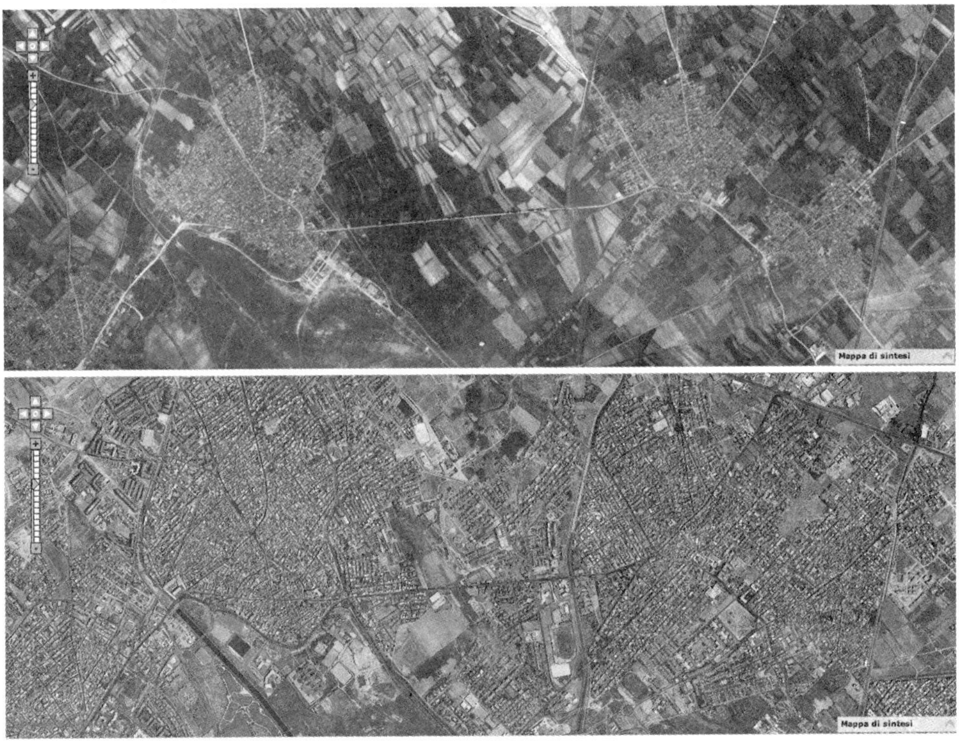

Figure 7 Urban development effects on the peri-urban cultivated land.

4.3 POLICY MEASURES

The Landscape Regional Plan acknowledges the weaknesses of the metropolitan green infrastructure, and suggests preserving its remaining elements. Those ones which Gilles Clement calls the "third landscape". Here, the void assumes the role of architecture stratified, which represents a cultural and collective value.

The protection of those spaces should encourage engineers and architects to use the peri-urban landscape as a generative element for the urban form. The compliance of the standard of 9 sqm/inhabitant of "green spaces", mostly delivered with flower beds and scattered gardens, could be better supported, for instance, by a park system with integrated management, the first pluralistic governance experience at an inter-municipal level.

At the local level we can notice two sets of territorial strategies attributable to the two main territorial subjects: the Municipality of Cagliari and the group of four towns that close the Gulf of Cagliari.

The capital, to counter the trend of movements towards the periphery and keep its population, has implementeda program of grants reserved for residents of the city who hold a lease for a principal residence, duly registered or filed with the Registry and is implementing the mutual exchange of accommodation, within the social housing program, between families too numerous, who live in houses too small, and small families that have unused rooms.

Recently the council has rejected the proposal for a large parcelling plan, located in the last whole green area on the north part of its territory, named in Sardinian language "Su Stangioni", which means "the pond", with a clear reference to its hydrogeological vulnerability.

The surrounding towns are planning their new development within the southern side of the highway SS 554 and are adopting some programs of urban regeneration to be completed with the availability of services and more residential buildings in those areas affected by illegal constructions, including the northern side of the SS 554.

It seems, however, that they are not working on a shared perspective for the "green areas", besides the admirable aims of the "Strategic Plan of the Wide Area of Cagliari". Actually, the problem that has pervaded the planning experience in those towns over the last 20 years lies on one important mistake: having been focused more on the urban expansion than on improving the quality of each historical central settlement to make it a better place to live in, attracting more population.

The original urban structure was formed by the addition of the different variants of the type of home court. It reflected a certain way of life very different from that of today, and that might be one of the reason why, instead of promoting an adaptation of that specific typology, it was endorsed to replace it with a greater volume of anonymous buildings. Those new buildings (two or three floors) have saturated the "gaps", typical of the courts, that kept the permeability of the urban fabric. It seems too dense, even claustrophobic.

5 SASSARI AND THE SUBURBAN BELT OF OLIVE TREES

Sassari provides one of the clearest examples of how the private-oriented development has significant effects on peri-urban agricultural landscapes.

Originally, Sassari developed through the slow concentration of scattered populations in the coastal regions, moving close together to defend themselves from pirates and avoid the precarious health conditions of low-lying areas leading to malaria.

In the mid Nineteenth Century, Sassari exhibits a flourishing economy based on agriculture, with a first ring around the village used for horticultural crops, and an outer one with specialized crops of vineyards and olive trees.

Vittorio Angius depicted Sassari with these words: "Sassari does not need to form a public garden, if its circumstances form a garden so varied, pleasant and vague, that art could not do better."

The Cominotti printing of 1827 (Figure 8) depicts the fields planted with olive trees around the city of Sassari, probably similar to the one that Vittorio Angius would see and assess three decades later. The most fascinating part in his observation concerns the recognition of the complementary roles of the countryside and the built city.

Figure 8 GiuseppeCominotti, View of Sassari, in direction of Porto Torres (1827)

As the city grows in a compact manner, a parallel "widespread settlement pattern" appears in the surrounding rural areas, characterized by villas that served as a second residence for the bourgeoisie of Sassari, as the roman villas had served the aristocracy in the past.

A first significant change in the green belt of olive trees landscape coincides with the implementation of the first petrochemical plants in the city of Porto Torres, 20 km North-West from Sassari in the 1960's. While the city was trying to give itself a shape through the regulatory instruments of the plan, it starts the phenomenon of building, as a result of an acceleration of the abandonment of traditional farming due to the perspective of a secure employment in the new industrial plant. New residential settlements, as Li Punti, San Giovanni and Ottava, have been developed next to the main road that connects Sassari and Porto Torres. Beside the effects on the morphology of the rural area, this process underlines a significant transition from an agriculture-centred economy to the industrial one.

Nowadays, the "grainy" city affects large parts of the rural landscape. A large portion of the population (72.68%) lives in the "compact and dense city", while the remaining population lives where scattered settlement prevails (Figure 9).

Working with the cadastral data of 1860 and 1920, professors Dettori and Filigheddu (University of Sassari) identified a higher density of olive groves in the first date on the ring around the city than it is at present. The plantations were associated with vineyards, pasture and arable land. From 1860 to 1920 the olive stands increased due to the destruction of the vineyards affected by grape Phylloxera, a disease imported from America. Between 1920 and 1977, the urban development caused the decline of the olive stands due to cover density reduction or land use change.

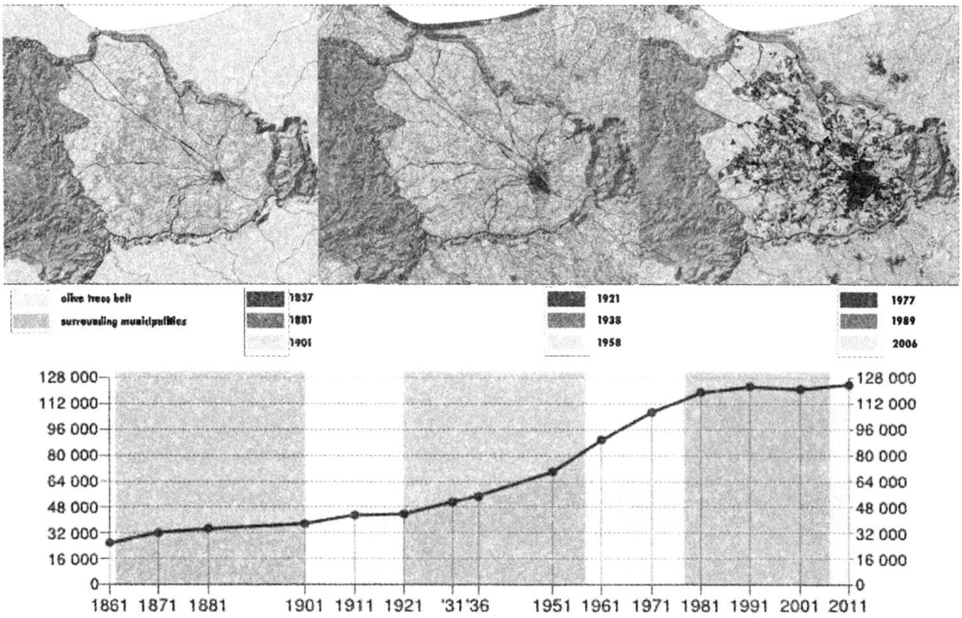

Figure 9 Stages of urban growth

Overall, the 57-year period sees the disappearance of approximately 134,000 olive trees (42 thousand for wear and 92 thousand for erosion), mostly located next to the city walls and, therefore, presumably old and of great environmental value and monumental. The expansion of crops in areas not in direct contact with the city has resulted in the planting of 36,000 trees for a final balance of the fifty years that ended with the loss of about 98,000 olive trees (Dettori and Filigheddu, 2008)

5.1 WHAT ARE THE FEATURES OF THESE DYNAMICS?

The "grainy city" (Figure 10), with single family homes fully equipped, a small isolated kingdom, represents a lifestyle suitable for family cell. (Figure 11)

Figure 10 The peri-urban "grainy" city

It is an "emporium of architectural styles" whose implementation is supported by the mass distribution of private vehicles, freezers for families and the development of communication technologies. Those who choose to live far from the chaos of the city are making their contribution to increase it.

Figure 11 The "cells"

5.2 POLICY MEASURES

The Sardinia Landscape Regional Plan states:

"The conservation of the relationship between land structure and the settlement should be based on maintaining the existing dimensional and volumetric ratios, in order to avoid an excessive fragmentation the land and to ensure, the maintenance of the landscape of olive groves carried out by the owners".

It also stated that the minimum area for building in the agricultural zone is 3 hectares, and the interventions must be connected with the rural activity; for other use the minimum area must be 5 hectares. In both cases the new constructions should be far away from protected coastal areas.

After a popular "protest", which ended with the election of a new regional government, new laws establish the minimum lot area for building in the agricultural zone to be 1 hectare, although it leaves to each single Municipality the decision to increase that minimum up to 3 hectares.

6 CONCLUSIONS

The traditional tools of planning practice and its implementation, as we can deduce from the two emblematic cases just told, process have failed. They have not been able to forecast social demands. But the biggest fault might lie in the choice to delegate the authoritative communication of the new common value of the landscape and soil protection to laws and abstract numbers.

The series of frescoes designed by Lorenzetti are composed by four scenes arranged along three walls of a rectangular room called the Hall of the Council of Nine, or of Peace, In Siena. The communicative power of the representation can convey an inspiring message into the work of the governors who gathered in those rooms to discuss the fate of their territories.

Facing the individual processes that have shaped spaces outside the dense city brings the need to identify an effective way to find and support a shared system of values, clearly identifiable through a concrete visual perspective, a human-shaped one.

The unconscious and the scenario are essential to preserve the value of the common good of the city and cooperative skills. The production of imaginary can be a ground for dialogue between experts and citizenship: if we do not find the "space" for dialogue, the city will always be more technical and corporate interests will prevail (Olmo, 2012). The distance between the expert knowledge and the community is exacerbated by languages, techniques and tools that are used to represent the development of the city. In 1764, when it takes really start the construction of the Church of Sainte Genevieve, then the Pantheon, in Paris, most of the time and skills (and were Soufflot and Perronet) decided to share what would have happened to citizenship by commissioning a famous painter, Robert, representation-on a scale of one to one-the portico of the church and the areas around (R. Gabetti and C. Olmo, 1989).

But it is not just a matter of tools, of course. The whole approach of planning has to be reviewed. The experience of the opposition made against the prescriptions of the Sardinian Landscape Regional Plan is illuminating: it tells us that there is kind of an on going territorial dispute, a disagreement over the possession/control of land between two or more territorial subjects. It is clear that this means, out of any prejudgments, that there are two or more different perspectives or conceptualizations on the future of the land (the environment and landscape) in the fringes of the two main Sardinian cities.

It is necessary to reassess public policies related to landscape stewardship, by shifting the priority from curbing the actual negative trends towards the achievement of shared quality objectives.

It has to be found a way to identify, even to represent the different perspectives: some might be translated as the final result of the additions of many individual ones; some other will be "holistic", but maybe too ideal if they are detached from the reality. The most successful way might start from recognizing that all of them should be based on the value of territory that is built on three foundations:

• the physical extension of terra firma is a good whose distribution matters to everyone (because we, our belongings and things we built individually and collectively, take up space)

• secure access to good land, land that we can use to do the things we care about, is essential to our capacity to make our way in the world (because land is made of resources that we need in order to survive, prosper and express ourselves)

• land and its properties (its location, its material composition, who or what lives on it) are essential to a vast array of world systems, such as nitrogen and carbon cycles, water purification and storage, ecosystems, and the production of oxygen, without which we would not exist (in our current form at least).

People have an interest in access to land that supports the sort of life that they lead. These lives are integrated with their geographic and ecological underpinnings.

Territory is both a highly particular good and a universal good. A theory of territory therefore must accommodate both the universal and the particular (Avery Kolers).

REFERENCES

[1] BERDINIP, SCANDURRA E, ATTILI G. Il Pianeta degli Urbanisti[J]. DeriveApprodi, 2013: 40-49.
[2] BOYLEJ. The public domain[M]. City of New Haven: Yale University Press, 2008.
[3] GONNER E C K. Common land and inclosure[M]. London: Macmillan&Co, 1912.

[4]HARVEY D. Spaces of global capitalism: a theory of uneven geographical development[M]. London: Verso, 2006.

[5]HOSKINS W G. The making of the English landscape[M]. London: Hodder and Stoughton, 1955.

[6]INDOVINA F. La metropolizzazione del territorio. Nuove gerarchie territoriali in economia e società regionale[J]. Oltre il Ponte, 2003:3-4.

[7]KOLERS A. Land, conflict and justice: a political theory of territory[M]. England: Cambridge University Press,2011.

[8]MATTEI U. Beni comuni. Un manifesto[M]. Laterza:Bari, 2011.

[9]MORLEY N. Metropolis and hinterland: the city of rome and the Italian economy[M]. England: Cambridge University Press, 2002.

[10]PIRASTU I. Banditismo in Sardegna[M]. Roma: Editori Riuniti, 1973.

[11]SECCHI B. La città dei ricchi e la città dei poveri, collana Anticorpi[M]. Bari: Laterza, 2013.

[12]STEEL C. Hungry city: how food shapes our lives[M]. UK: Random House UK Ltd, 2013.

[13]OLMO C. Viaggio nei comuni delle buone pratiche[R]. Torino: Biennale spazio pubblico, 2013.

Part 4 Cold Climate Regions: Challenging Context to Experiment Low Carbon Target Development

On the "Low-carbon" Approach to Cope with Weather Vulnerability: A Case Study of the Snow Village of China

Wen Cheng, Tianyu Zhao

Harbin Institute of Technology, Department of Architecture, Heilongjiang, Harbin

ABSTRACT: The rapid development of tourism has brought many benefits to the Snow Village, but with the influx of a large number of tourists, the fragile environment has been gradually affected. High carbon emissions originated from tourism activities even impact climate environment in local district. After years of tracking, surveying and research it has been figured out that the main problem is the massive influx of tourists, concentrated over the capacity, the high number of motor vehicles, the overuse of traditional energy sources, the improper sewage discharge, rubbish and so on. Therefore, aiming to achieve the goal of low-carbon and make the development of snow village ecologically sustainable, this paper focuses on research and implementation of feasible planning strategies, dealing with the layout of the village, its traffic patterns, energy, sewage and waste discharges, etc. to provide a base for the development needs and the local climate characteristics.

Key words: The snow village of China, climate vulnerability, low carbon, low carbon planning

1 INTRODUCTION

In recent years, climate changes and related scientific issues have drawn increasing attention in the academic circles, and the climate warming that results from carbon emissions has become one of the hottest issues. In 2007, The Fourth Assessment Report of Intergovernmental Panel on Climate Change (IPCC) demonstrated the correlation between carbon emissions and climate, warning through a wide set of scientific data that: "The rise of the average global temperature is due in some degree (The possibility is believed to be over 90% according to some forecast) to the increased concentrations of greenhouse gases which are emitted by human activities" (page 10), thus the low carbonization development becomes an inevitable choice to curb global warming. Therefore, as an important method to guide social development, integrate environmental resources and regulate human activities, all levels of planning work should play an important role in reducing greenhouse gas emissions from human activities, and become an efficient allocation mechanism of "low carbon" action plan.

2. SNOW VILLAGE OF CHINA AND ITS CLIMATE VULNERABILITY

2.1 SNOW VILLAGE

The Snow Village of China is located in Shuang-Feng Forest of Dahailin Forestry administration in Heilongjiang Province. Due to the influence of regional geographical environment, the cold air from north Siberian Baikal Lake of Russia and the flow of warm moist ocean air from south Japan intersect here frequently, and produce the 5-month-long snow. Its wonderful snow shape and the temperature, humidity, wind and other weather conditions are dependent on climate changes and its environment sensitivity has created a different snow landscape. It is also famous for its well-known snow scenery, snow rhyme and snow sculpture which are carved naturally by wind (Figure 1).

Figure 1 Graceful snow scenery of Snow Village

Tourism to the Snow Village is increasing at high speed and rough-type development has taken place, besides with the increasing number of visitors and the rapid expansion of the construction reception scale (Figure 2), related infrastructure and planning management do not meet the demand. Especially during the peak travel period in winter, the small mountain village suffers from unsatisfactory phenomena, such as overcrowded (Table 1). This kind of disorder construction and excess emissions not only changes the original climate pattern but also has a profound effect on the locally fragile air temperature, humidity, wind direction and other related weather conditions.

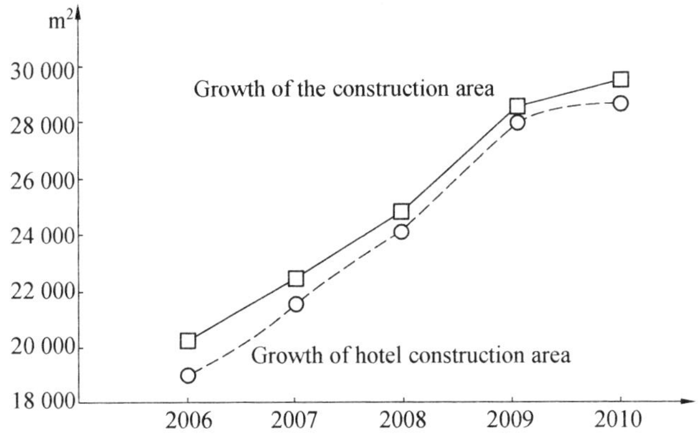

Figure 2 Growth of the construction area

Table 1 Visitors quantity growth of Snow Village in nearly years (Author 2013)

Year / Visitors quantity	2006	2007	2008	2009	2010	2011	2012
Annual total	41 230	54 882	80 120	102 700	112 420	178 015	230 123
Winter tour stage	39 660	51 782	71 370	87 520	93 400	153 030	168 022
Post/pre-winter season	1 570	3 090	8 750	15 180	19 020	24 985	62 103
Visitors quantity/peak ten days	20 210	27 211	38 090	41 400	37 600	27 802	31 400
Visitors quantity / peak day	3 200	3 630	4 110	5 100	4 210	3 899	4 725

2.2 THE CLIMATE VULNERABILITY

The natural system in Snow Village is a typically fragile one, including local temperature, humidity, wind direction, wind speed, the surface shallow water streams, sensitive alpine vegetation, the pure air, etc. It has a high possibility to be interfered and damaged by external factors, by both direct and indirect impacts of human activities and vehicle emissions, and shows less competence to the negative impacts of climate change.

The climate vulnerability of Snow Village is an outcome of its geographical conditions as well as external influences like human activities and greenhouse gas emissions. According to what has been observed during the winter tourism season, the average inner temperature in Snow Village during the daytime is 2 degree higher than the one in its surroundings, and the annual increase has been of 0.3 degrees during the last decade. The temperature rise in the microclimate of the valley leads to local snow change and reduction, wind speed and direction change, affecting the shape of the snow landscape, increasing the presence of smoke and dust pollutes on the white snow. These changes are all damaging the local ecological landscape and environment. Although global climate change is one factor, a more clear reason appears to be the overloaded tourism activities that disturbs local climate.

3 THE TOURISM ACTIVITIES IMPACT OF HIGH CARBON EMISSION TO LOCAL CIRCUMSTANCE

There are number of external factors which threaten the climate environment of Snow Village. In order to study the environmental harm factors further, surveys were made respectively in 2008, 2010 and 2012 and focused on some prominent environment impact problems of the winter tourism peak. Through the survey we found that things such as foreign visitors' traffic activities, consumption demands, waste products, construction activities and improper recreation behaviours produce high peaks of energy consumption, high emissions and high pollution as well as influencing factors of climate environment.

3.1 THE HIGH CARBON EMISSIONS OF EXTERNAL VEHICLE

Located in a mountain area, Snow Village can only be reached by vehicles. With the development of transport infrastructure, the original fleet typically composed by small and medium-sized station wagon cars has been gradually replaced by sedan cars and large SUVs, among which the private cars' traffic volume is rising rapidly and accounts for approximately 70% of the total (Table 2). Cars need drive 30 kilometers along through the valley area of Hailang River which is the most sensitive part of the region to reach the core recreation Shuangfeng Forest. However this recreational area is only a small village which covers 28 square kilometers, with the narrow roads and increasing cars both of dynamic traffic and static traffic bring chaos (Figure 3).

Table 2 Vehicles amount of Snow Village in nearly years (Author 2013)

Year / Visitors quantity	2006	2007	2008	2009	2010	2011	2012
Annual total	2 820	3 460	4 670	6 460	6 780	—	—
Winter tour stage	2 010	2 370	3 570	5 190	5 680	7 400	11 200
Visitors quantity/peak ten days	890	1 250	1 900	2 680	2 760	763	1 250
Visitors quantity/peak day	210	273	310	329	380	172	240

Figure 3 Transportation exhaust and pollution situation in Snow Village

Congestion often appears during the peak travel period which not only leads to massive emissions and heat flow towards the micro-environment, but also impacts local climate, temperature and snow quality.

3.2 THE HIGH CARBON EMISSIONS OF TOURISTS

With the increasing visitors, demands related to consumption such as food, housing, heating, excretion, just to mention a few, are at the height of tourist season nearly 30 times higher than that of original residents (according to the estimation of the local administrators). And apart from that, per capital carbon emissions of visitors result from food, waste product and heating are 1.5 times as much as that of local villagers (estimated mainly according to the answers of the local residents about food, supplies, water and electricity consumption, burn wood consumption, etc.). Daily life rubbish causes pressure and disturbance to primitive infrastructure and environment, while sewage is discharged onto the Hailang River and both sides of the riverbank, which leads to directly environmental pollution (Figure 4).

Figure 4 Garbage on roadside

3.3 THE HIGH CARBON EMISSIONS OF CONSTRUCTION ACTIVITIES

In order to meet the increase of tourism demand and uplift the service reception level, construction activities in Snow Village are hyperactive in these years. Annual new expansion amount in 2006, 2007, 2008 is more than 2,000 square meters (most of them are single-storey and double-storey buildings) and the amount expands to 6,900 square meters in January 2009. The tourist industry of Snow Village started from 2000, and until now the construction land of Shuang Feng Forest, which is a core scenic spot, has increased by 67%, making the building junk and building energy the major source of high carbon emissions.

3.4 THE HIGH CARBON EMISSIONS OF ENTERTAINMENT ACTIVITIES

With the gradual development of tourism management, entertainment projects provided by Snow Village are much more plentiful than before. Therefore, some high carbon emissions projects like fireworks, firecrackers and snowmobiles continue to persist, due to the considerable income. Particularly, a large number of additional carbon emissions and dust pollution are caused by firecrackers every evening at the peak period of Spring Festival.

4 PLANNING APPROACH TO ACHIEVE A LOW CARBON TARGET IN SNOW VILLAGE

Planning can produce a long-term structural adjustment and is able to control the development of a region and has also impacts to its socio-economic activities. By comprehensive planning, integration of transportation, standard construction practices such as planning approach, Snow Village is planning to raise the adaptability of the environment and tourism, focusing on their vital landscape conservation, and realize its target of tourism development of low carbon and low interference with the climate.

4.1 BALANCING PLANNING AND IMPROVING LOCAL CARBON ENVIRONMENT

Now the capacity of Snow Village Shuang-Feng forest farm environment has been seriously overloaded, in urge of increase, share capacity, and the adjacent Yong-An forest farm, Bai-Song forest farm and Film Base all have similar basic conditions for the landscape, located in the inevitable way towards Snow Village.

The original single-point development approach is planned to extend to a cluster planning as "four district", "one valley" and "seven spots" (Figure 5). On the one hand, the layout of the tourism areas will be adjusted in the plan, on the other the growing tourism demand will be dispersed. Furthermore, the expansion construction of the core area should be controlled on capacity and carbon emissions of the area, while promoting changes in land use intensification (Figure 6), balanced distribution of visitors, and using carbon sinks to improve the conditions of spatial scales are also suggested.

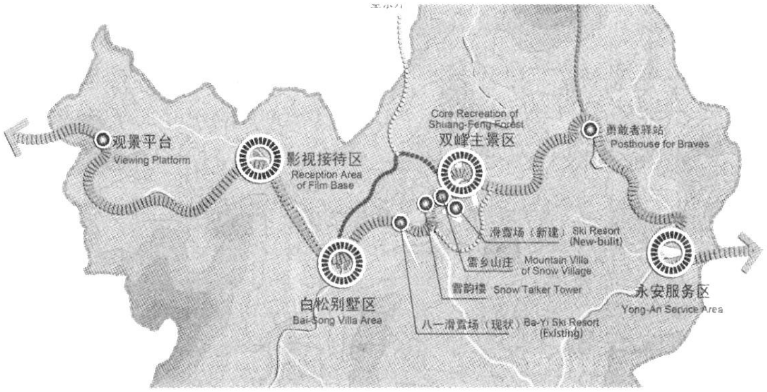

Figure 5 The macro-layout and decentralize press

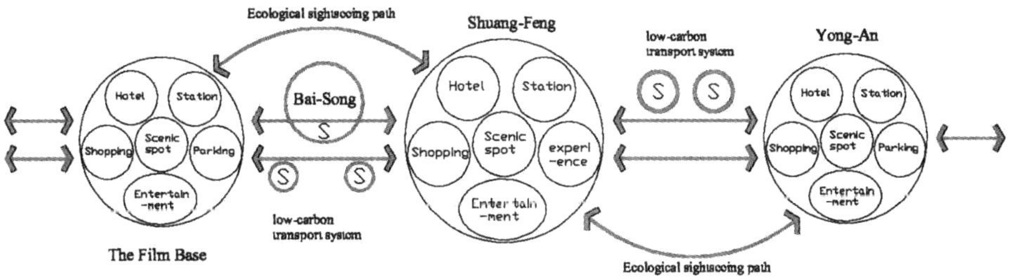

Figure 6 Schemes of layout and low-carbon transport system

4.2 INTEGRATING TRAFFIC STRUCTURE AND REDUCING CARBON EMISSIONS

To ease the traffic pressure at the core recreation spot of Shuang-Feng Forest and control a variety of tourist vehicles entering ecologically sensitive parts of the area, it is necessary to organize tourism traffic planning in a wide range, build parking lot in Yong-An forest farms and the Film Base as there are more adequate free space. Then the entire 24km climate vulnerable zone as non-carbon transportation area should be set from Yong-An forest farm to the Film Base (Shuang-Feng is in the middle). Still various types of low-emission means of transport are fit for the micro-climate vulnerability region, for instant, tourist buses, small tram, solar car, dog pulled sledge, deer pulled sledge, carriage and so on. Finally a low-carbon transport system should be created with both enjoyment and practicability by taking more complex traffic means like ice & land bi-track road.

4. 3 PROMOTING COMPOSITE APPLICATION OF LOW TECHNOLOGY AND LOW CARBON TECHNOLOGIES

Building is an important element of the snow landscape in Snow Valley with unique forest features. It reflects the simplicity of this remote mountain village. It is simple, beautiful, and environmental-friendly. In order to create unique low carbon-emission buildings in Snow Village, when renovating and upgrading those architectures, it is better to focus on energy efficiency based on local and traditional features, encourage the use of eco-low-technology and new composite materials, new low-carbon technologies, such as hybrid hi-tech compound wall, ground heat pump technology, insulation technology with the shape, material and ground heat of local buildings.

4.4 PLANNING THE LOW-CARBON INFRASTRUCTURAL FACILITIES

As an aim to change the energy structure of the whole scenic spot and reduce carbon emissions, a long-term plan should be carried out to make full use of the wind energy and establish a small wind power. For instance, in the west side of the Snow Village, the conditions for wind energy are really favourable, with the height of 50 meters and the average wind speed of 6. 9 m/s, although the construction on a challenging topography should be overcome.

Another important goal is to classify the rubbish of the whole scenic spot and increase the frequency of collecting and transporting. The rubbish disposal plant is planned to be established in the leeward area of Yong-An tree farm, keeping the dumping ground under strict supervision. It is suggested to call on the users to set the simple basic water circling system and small ecological sewage treatment plants will be established for every scenic spot to reduce water consumption and sewage disposal.

4.5 ADVOCATING LOW CARBON TOURIST ENTERTAINMENT ACTIVITIES

In order to promote low carbon tourist entertainment activities, reduce air pollution and provide security to tourists & houses (using timber walls), it is suggested to focus on several strategies: promoting actively, strengthening the management, using various health and unique entertainments to take the place of high carbon-emissions activities, forbidding sale and display of fireworks, moreover, gradually reducing the use of high oil-consumption snowmobiles in the core area.

5 CONCLUSION

Low-carbon development is the inexorable choice for the Snow Village to cope with the environmental crisis and stunt the temperature increasing. Base on this, it is necessary to plan reasonable layout and land use, adjust the structure of traffic and transportation, make program for using clean ecological energy and build practical and economic infrastructural facilities, etc. Furthermore an effective comprehensive configuration mechanism can be formed. All these actions could make the planning as an effective way to implement the regional low-carbon development, to conform the resources and environment, to improve and defining people's behavior.

REFERENCES

[1] GU Chaolin, TAN Zongbo, LIU Zhilin. A possible approach of urban planning for low-carbon city [J]. Journal of Urban and Regional Planning, 2010, 3(2): 23-42.

[2] CHENG Wen, ZHAO Tianyu. Study on prevention of eco-landscape risk in scenic area[J]. International Disaster and Risk Conference, IDRC, 2010,11: 309-315.

[3] FUSSEL H M. Vulnerability: a generally applicable conceptual framework for climate change research[C]. Global Environmental Change, 2007, 7(2): 155-167.

[4] Dahailin Forestry Administration. The construction planning of the snow village national park (the Shuang-Feng Forest Farm). The Explanatory Memorandum, 2007: 3-7.

[5] Dahailin Forestry Administration. Research and planning of tourism scenic spot of China's snow village. The Explanatory Memorandum, 2009, 3-8.

[6] STANLEY C T Y. Low carbon city planning: carbon sink assessment model for urban-rural ecological green space system. city planning review[J]. The Journal of UPSC, 2011, 35(2): 32-38.

[7] JIN Hong, LING Wei. Low energy consumption, low technology, low cost—research on energy-saving house design in winter tow[J]. Architectural Journal, 2010, 8: 14-16.

Dynamic Analysis for Solar Energy Availability in Harbin High-rise Residential Area

Ming Lu, Jun Xing, Shuxin Chen
School of Architecture, Harbin Institute of Technology, Harbin 150006, *China*
hitlm@ 126. *com*

ABSTRACT: The amount of solar radiation captured directly affects the solar energy utilization efficiency of city's building groups, and then affects the energy saving in urban residential area. So it's necessary to simulate and analyse the amount of solar radiation obtained in residential area. This contribution firstly investigates the high-rise residential area newly constructed in Harbin from 2008 to 2013. Two typical layout models are selected as the objects of simulation and analysis. They are 18-floor parallel determinant layout model and 18-floor three sides surrounded layout model. Then the paper summarizes the simulation's results of the solar availability of four high-rise residential building's facades in winter by using the light environment simulation software Ecotect and the statistical analysis software SPSS, and analyses them. Finally, it puts forward the optimal strategies of the two layout models based on three aspects: layout orientation, vertical wall distance and gable wall distance.

Key Words: Harbin, high-rise residential area, solar radiation availability, layout optimization

1 INTRODUCTION

According to the study, the annual reserves of solar energy reached 1.7 trillion t of standard coal in China. It can be comparable to tens of thousands of generating capacity of the Three Gorges Project. China has begun to enter the new urban era with urban-based society playing the dominant role. The increase of urban building density makes a lot of buildings severely shadowed. Many solar technologies can not get the expected profits. And the contradiction between high-density urban fabric and utilization of solar energy is increasing. As we know, more and more researches on the urban sustainable development have been focused on energy use. Being a kind of green and renewable energy resource, solar energy utilization has been paid more and more attention.

A European research shows that urban planning layout has a great influence on the amount of solar radiation captured. Different layout patterns affect the mutual shelter between buildings, and then the range of the amount of solar radiation obtained on building surfaces is up to 20% ~ 80%. Thus, the layout of building groups can have a direct impact on the efficient utilization of solar energy. We know that the most fundamental condition for solar energy utilization is that buildings must be fully exposed to the sun in order to ensure a certain amount of solar energy availability. But with the urban development, the building density is getting higher and higher, especially in the residential area, creating more shadows among buildings, and greatly affecting the solar availability.

Harbin is a winter city located in the North-east of China. Coal is commonly used in winter as fuel for heating and it represents one of the most influent causes of air pollution. Together with the low sunshine amount and long heating cycle in the North of China caused by the impact of the natural environment latitude, it is urgent to adopt a new energy-saving policy and an environmentally sustainable resource as the alternative fuel. It is very necessary to utilize the solar energy as much as possible in winter. Therefore, the research investigated the solar radiation conditions of Harbin high-rise residential area in winter, and the way to optimize the solar radiation. It would provide a basic reference to improve the local regulations and basic research on energy conservation of city building groups.

2 BASIC CONDITIONS OF SOLAR RADIATION SIMULATION

2.1 BASIC SITUATION OF SOLAR RADIATION IN HARBIN

The distribution of solar energy resources in Heilongjiang Province is located in the Ⅲ zone in which the amount of solar radiation captured is on the average level of China. The total average solar radiation captured in a year in Harbin was 4 200 ~ 6 100 MJ/m^2. It has the potential for the development of solar technology. According to the investigation from China Meteorological Administration, the average of annual solar radiation in Harbin is about 4 718.7 MJ/m^2. It means that Harbin represents a rich area of solar radiation in China. So it is very suitable for the utilization of solar energy resources and the comprehensive development. Therefore, it's very important to get high amount of solar radiation during the utilization process of solar energy.

2.2 LAYOUT MODELS AND SIMULATIVE OBJECTS

SPSS (Statistical Product and Service Solutions) has been used to summarize the research data of 68 high-rise settlements from 2008 to 2013 in Harbin, and then contrast and analyse the data.

Because of the highest proportion in layouts and floors, two typical layout models are selected as the simulation objects. Model A is 18-floor parallel determinant layout. Model B is 18-floor three sides surrounded layout. The two typical layout models are shown in Figure 1.

Four building facades (south, north, west and east) of the building # 05-18FPR in parallel determinant layout model and build # 03-18FTHS in three sides surrounded layout model are selected as the simulative facades. The west, middle and east points on the centre height of each floor are chosen as the simulative points. The details are shown in Figure 2.

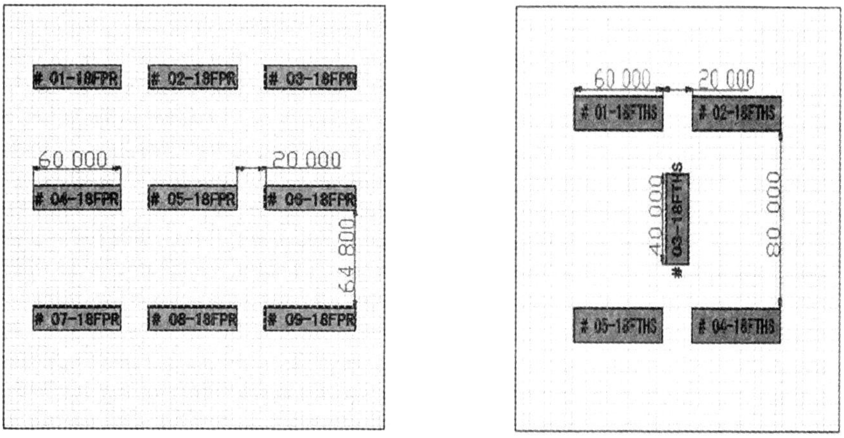

(1) Model A (2) Model B

Figure 1 Layout and dimension of two typical models

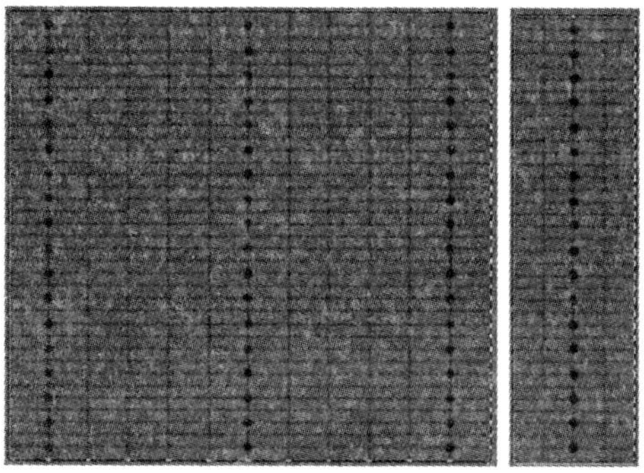

Figure 2 Simulative points on building facades

2.3 BASIC SIMULATION CONDITIONS

This study presented a simulation analysis by using light environment analysis software Ecotect and statistical analysis software SPSS.

Solar radiation reaches the ground through the atmosphere after weakening. The incident solar radiation on building facades includes two parts: direct and diffuse radiation. In this paper, the sum of the two part solar radiation (unit: Wh) defines the size of the amount of solar radiation obtained. Harbin meteorological data is used in this paper (latitude 45.7, longitude 126.7, time zone, +8). In the simulation analysis, the cumulative value is selected. Therefore, we conducted the simulation of solar radiation on a quarterly basis. Specific simulative time period is: spring (March to May), summer (June to August), autumn (September to November) and winter (December to next February).

3 SIMULATION AND ANALYSIS

This study has analysed the amount of solar radiation obtained on the four building facades in winter which is of the 18F parallel determinant and three sides surrounded layout models by using Ecotect. It gets the special value and average of solar radiation, and then conducts the comparative analysis of these values.

3.1 SIMULATION AND ANALYSIS OF MODEL A

3.1.1 THE SOLAR RADIATION ANALYSIS OF FOUR FACADES

Table 1 gives the special values and average values of solar radiation of each facade in winter in parallel determinant layout.

Table 1 Special and average values of solar radiation on each facade in Model A

Building facade	South	North	Western	East
Maximum/(Wh)	152 569	46 643	98 259	101 018
Maximum position	East point on 17F	West point on 17F	18F	18F
Minimum/(Wh)	55 489	42 230	60 187	52 969
Minimum position	Central Point on 01F	Central Point on 01F	01F	01F
Average/(Wh)	112 117	45 085	82 314	81 916
Over Average position	West point on 10F	West point on 09F	10F	10F

As shown in the Table 1, the south wall can get more solar radiation than others. The position of east point on 17F can get the maximum solar radiation. On the contrary, the north wall gets far less solar radiation than the others. The position of central point on 01F can get the minimum solar radiation. There is a little difference between the east and west wall. So the south wall is the best area for solar utilization.

3.1.2 ANALYSIS OF THE FACADE WITH OPTIMAL SOLAR RADIATION

Figure 3 illustrates the solar radiation of the left, middle and right position on each floor of south wall in winter. They have the similar upward trend. And the curve of each simulative point has the similar rising laws. On the height interval of 01F ~ 03F, changes of solar radiation is relatively stable. On the height interval of 03F ~ 14F, the solar radiation has been in a rapid upswing. And on the 10th floor, the solar radiation of each simulative point has been higher than the average value. On the height interval of 14F ~ 18F, the solar radiation is in the steady interval after rising. And it is at the maximum with floating upwards and downwards slightly.

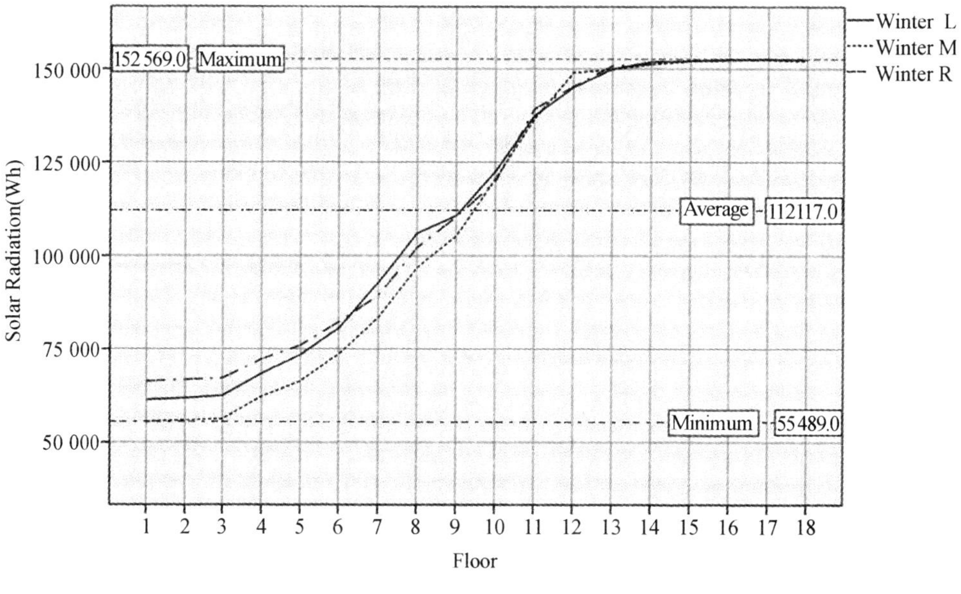

Figure 3 Changes of solar radiation on south facade in parallel determinant

3.2 SIMULATION AND ANALYSIS OF MODEL B

3.2.1 THE SOLAR RADIATION ANALYSIS OF FOUR FACADES

Table 2 gives the special values and average values of solar radiation of each facade in winter in three sides surrounded layout.

Table 2 Special and average values of solar radiation on each facade in Model B

Building facade	South	North	West	East
Maximum(Wh)	152 860	46 589	98 310	101 023
Maximum position	18F	18F	West point on 18F	East point on 18F
Minimum(Wh)	73 625	32 598	39 497	36 181
Minimum position	01F	01F	East point on 01F	West point on 01F
Average(Wh)	92 966	39 658	64 682	61 331
Over Average position	14F	10F	East point on 09F	East point on 10F

A sit can be seen from this table, the simulation result is similar with Model A. the south wall can get more solar radiation than the others. On the contrary, the north wall gets less solar radiation than the others. There is a little difference between the east and west wall. So the south wall is also the best area for solar utilization.

3.2.2 ANALYSIS OF THE FACADE WITH OPTIMAL SOLAR RADIATION

Figure 4 illustrates the solar radiation of south wall in winter. It is an upward trend. On the height interval of 01F ~ 12F, the solar radiation has been in a slow upswing. On the height interval of 12F ~17F, the solar radiation has risen rapidly. And on the 14th floor, the solar radiation of each simulative point has been higher than the average value. On the height interval of 17F ~18F, the solar radiation is in the steady interval after rising.

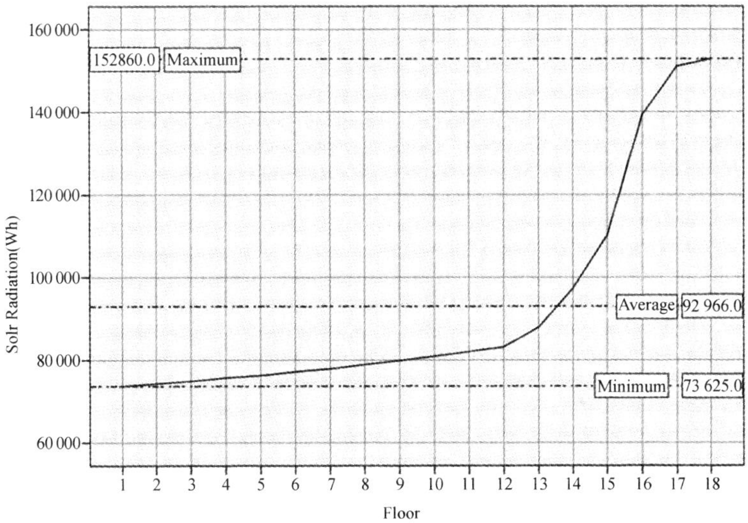

Figure 4 Changes of solar radiation on south facade in three sides surrounded layout

4 OPTIMIZATION AND EVALUATION

4.1 OPTIMIZATION AND EVALUATION OF MODEL A

4.1.1 BUILDING ORIENTATION

With the layout model unchanged, this study mainly carries on the simulation analysis of solar radiation in winter with simulative orientations changing every 15° in parallel determinant layout model. The south facade of the building #05-18FPR is adopted as the object of study. The average is calculated by the results of the simulation analysis showed above. Then the study compares the corresponding average of each orientation with the average of south facade. Table 3 gives the average values of each orientation, which are higher than the south one.

Table 3 Average values of each orientation

Layout orientation	-60°	-45°	-30°	-15°	0°	30°	45°	60°
Average/(Wh)	126 061	127 703	118 951	117 983	112 117	121 055	123 600	118 030
Percent of increase	6.1%	13.9%	12.4%	5.2%	0%	8.0%	10.2%	5.3%

As can be seen from Table 3, for the solar radiation in winter, south-west orientation is better than the south-east orientation on the whole. The percentage of increased average first increases and then decreases from 0° orientation to 60° orientation. The average of 45° south-west orientation increased the most compared with south orientation. And the 45° south-east orientation increased the lowest.

4.1.2 VERTICAL WALL DISTANCE

In this study, the vertical wall distances of 1.4H, 1.6H, 1.8H, 2.0H and 2.2H were adopted to compare with 1.2H which is the minimum vertical wall distance by an urban planning code (Regulation of Urban and Rural Planning in Harbin). Table 4 gives the solar radiation in winter and the FAR of the south facade in different vertical wall distances. According to Regulation of Urban and Rural Planning in Harbin, this study adopts 10 m building back the red line distance, and it assumes that roads surrounding simulation blocks are city branch level roads when calculating the FAR.

Table 4 Comparative analysis of solar radiation in different vertical wall distances

Vertical wall distance	1.2H	1.4H	1.6H	1.8H	2.0H	2.2H
	64.8 m	75.6 m	86.4 m	97.2 m	108 m	120 m
Average(Wh)	112 117	121 657	130 755	136 840	141 226	145 218
Percentage of increase	0.00	8.50%	16.60%	22.1%	26.0%	29.5%
FAR(Floor area ratio)	3.12	2.81	2.55	2.34	2.16	1.99
Percentage of decrease	0.00	9.94%	18.3%	25.0%	30.8%	36.2%
Ratio of increase to decrease	0.00	0.86	0.91	0.88	0.84	0.81

As can be seen from Table 4, with the vertical wall distance increasing, the average of solar radiation increases, and the FAR decreases. The increasing of vertical wall distance will inevitably lead to a decline in FAR. FAR is an important indicator to measure the construction land use strength. Therefore, considering the efficient use of land and energy saving, this study should analyse the optimal distance of solar radiation through the ratio of increase to decrease. When the FAR is constant, the more the average increases, the higher the ratio is. When the average is constant, the less the FAR decreases, the higher the ratio is. Therefore, the higher the ratio is, the more solar radiation availability will be got. From here we see that distance coefficients of vertical wall from best to worst are 1.6, 1.8, 2, 1.4 and 2.2. So we can see that 1.6H is the best distance for land use and solar availability.

4.2 OPTIMIZATION AND EVALUATION OF MODEL B

4.2.1 BUILDING ORIENTATION

By the same way of model A, with simulative orientations changing every 15° in three sides surrounded building #03-18FGER, the simulating result can be got. Similarly, the south facade of the building #03-18FGER is taken as the object of study. Table 5 gives the average values of each orientation, which are higher than the south one.

Table 5 Average values of each orientation

Layout orientation	-60°	-45°	-30°	-15°	0°	30°	45°	60°
Average/(Wh)	122 110	133 022	133 837	126 371	116 148	127 626	129 898	127 626
Percent of increase	5.1%	14.5%	15.2%	8.8%	0%	9.9%	11.8%	9.9%

As can be seen from Table 5, the average of solar radiation when facing south orientation is the highest. And the average of solar radiation when facing 15° southeast orientation is the most close to the south orientation.

4.2.2 GABLE WALL DISTANCE

In this study, increasing the distance until 10m in an interval of 2 m optimizes the gable wall distance. And then the optimized distances were compared with 20 m which is the minimum gable wall distance by an urban planning code (Regulation of Urban and Rural Planning in Harbin). Table 6 gives the solar radiation in winter and FAR of the south façade in different gable wall distances. The calculation method of FAR is the same as above.

Table 6 Comparative analysis of solar radiation in different gable wall distances

Gable wall distance	20 m	22 m	24 m	26 m	28 m	30 m
Average/(Wh)	92 966	96 787	99 622	100 980	104 417	109 419
Percent of increase		4.1%	7.2%	8.6%	12.3%	17.7%
FAR(Floor area ratio)	3.89	3.85	3.80	3.75	3.71	3.67
Percent of decrease		1.03%	2.31%	3.60%	4.63%	5.66%
Ratio of increase to decrease		3.98	3.12	2.39	2.66	3.13

As can be seen from Table 6, with the gable wall distance increasing, the average of solar radiation increases, and the FAR decreases. The increasing of vertical wall distance will inevitably lead to a decline in FAR. FAR is an important indicator to measure the construction land use strength. Therefore, considering the efficient use of land and energy saving, this study should analyse the optimal distance of solar radiation through the ratio of increase to decrease. When the FAR is constant, the more the average increases, the higher the ratio is. When the average is constant, the less the FAR decreases, the higher the ratio is. Therefore, the higher the ratio is, the more beneficial to utilization of solar radiation it is. From here we see that distances of the gable wall from best to worst are respectively 22 m, 30 m, 24 m, 28 m and 26 m.

5 CONCLUSIONS

This study has investigated the effect on solar radiation availability of orientations, building layouts and positions at facades in high-density residential areas through Ecotect and SPSS. The main findings are following:

- in terms of solar energy availability between two models, little difference of variation trend on every facade of building are comprehensively found;

- for each facade of two models, significant difference of solar energy availability can be found. The south wall is the best area, on the contrary, the north wall is the worst area. On the south facade in two models, they all have the upward trend from 01floor to 18 floor, but rate of rise on different floor changes greatly;

- the orientations take more effects on two models, in particular for the winter period. South-west orientation is better than the south-east orientation on the whole. The percentage of increased average first increases and then decreases from 0° orientation to 60° orientation. The average of 45°south-west orientation increased the most compared with south orientation. And the 45°south-east orientation increased the lowest;

- the vertical wall distance also take more effects on two models in winter. Considering the efficient use of land and energy saving, 1.6H is the best distance for model A and 22 m is the best distance for model B.

These conclusions are obviously limited to the specific building geometries, layouts and climate conditions. Several more complicated building layouts and a larger-scale urban area will be the subject of the work in the next stage.

ACKNOWLEDGEMENTS

This investigation was supported by the project of Harbin Innovative Talents of Science and Technology (Dynamic Analysis and Effective Utilization of Solar Radiation in Harbin High-density residential area, China; No. 2013RFLXJ028). This paper is one of the rescarch results of solar radiation in living areas in Harbin.

REFERENCES

[1] CHENG V, STEEMERS K, MONTAVON M, et al. Urban form, density and solar potential[C]. Proceedings of the 23rd Conference on Passive and Low Energy Architecture. Geneva, Switzerland, 2006,9:6-8.

[2] ZHOU Chuanrui. Assessment andanalysis of development and utilization prospects of solar energy resource in Harbin area [D]. Gansu: Lanzhou University,2012.

[3] ROBINSON D, STONE A. Solar radiation modelling in the urban context[J]. Solar Energy, 2004,77:205-309.

[4] LU Weide. Comment on the development trend of solar utilization technology[J]. World Sci-tech

R&D,2007,2:95-99.
[5]ZHAO Jilong ,TANG Yifeng. Integration construction of city planning and solar technology[J]. Construction and Architecture,2010,23:10-11.
[6]SANTAMOURIS M, PAPANIKOLAOU N, LIVADA I et al. On the impact of urban climate on the energy consumption of buildings[J]. Solar Energy, 2007,70:201-216.

Thinking about the Urban Development in Northeast of China in the Background of Economic Transition

Yu Zhang[1,a], *Ziguang Chen*[2]

[1]*School of Architecture, Harbin Institute of Technology, China, No.* 66 *Xidazhi Street, Harbin*, 150006

[2]*Architectural Engineering School, Heilongjiang University, China, No.* 52 *Xuefu Road, Harbin*, 150080

[a]*yu. zhang@ hit. edu. cn*

ABSTRACT: China is going through a period of economic transition which is aimed at keeping rapid speed and optimizing development mode as high efficiency development mode. The region of the northeast part of China used to be the national industrial center and it is facing the transition problem in this background. In the past years, there had been a lot of problems made by urban expansion. Present study analyzes the characteristics of city in northeast part of China in this special period, with regard to the local climate factors, summarizes the trends of urban development and proposes an urban planning strategy for city's development in this area of China in the background of economic transition.

Key words: cold climate city, low carbon development, economic transition

1 INTRODUCTION

With the high speed economic and social development in China, the national economy had been raised a lot in recent years. In 2011, the GDP (Gross Domestic Product) transcend that of America and became the No. 1 in the world. The central government fixed the aim of GDP growth as 7% every year. If China wants to keep such rapid GDP growth, it has to face two contradictions: one is the limitation of raw material and the industrial structure based on high not-renewable energy consumption; the other is limited agriculture land and huge land supply demand caused by high speed urbanization process. These two contradictions are more serious in northeast part of China where the weather conditions are very severe during winter and require a long time heating so that the energy consumption in building sector is higher than in other Chinese regions. On the other hand, as the traditional industry in China, the industry in northeast part of China is that of high energy consumption and high pollution. For example, in 2013, heavy industry, including energy, petrochemical and equipment manufacturing takes 70% in the whole industrial added value, while that of Jilin province and Liaoning province take 52.8% and 63%. Therefore, driving the low carbon and sustainable urban development is a significant topic for northeast part of China. In the same time, the urban development strategy for northeast part of China with cold climate is a significant research topic.

2 BASIC CONCEPTION ANALYSIS

2.1 LOCAL GEOGRAPHY FEATURES ANALYSIS IN NORTHEAST

The concept of "cold climate city" is not a very clear one: being formulated according to the climate in winter of the location of the city, the characteristics of the cold climate city refer to long-time winter season, lower temperature the whole year and high latitude area.

China is a country with a vast territory and there is a huge difference between climate characteristics of different regions. The northeast part of China which the predominant wind direction in winter are north wind and northerly wind locates in high latitude area and the period of low temperature last long time, this area is a typical cold climate area with typical cold climate cities(Table 1).

Table 1 Data collection of northeast part of China

	Area/ (10^3 sqm)	Population/ (10^3 sqm)	GDP/ (10^3 yuan)	GDP per capita/ (10^3yuan)	Lowest temperature (Capital)/(℃)	Humidity /(%)
National	960	134 735	4 715 640 000	3.508 3	-27.4	56.8
Heilongjiang	45.48	3 831	10 088	3.263 7	-38.1	65.4
Liaoning	14.59	4 375	17 500	5.034 9	-32.9	63.2
Jilin	18.74	2 746	8 460	3.834 7	-35	63.4

Data source: NBS (National Bureau of Statistics) (2011) and State Meteorological Administration

In the meanwhile, this part refers to Heilongjiang, Jilin and Liaoning provinces which were the old industry base of China. With the economic development of China, the central government begins to adjust the economic model in targeted region to act in cooperation with the national carbon emission reduction plan. The aim of the economic adjustment is to develop new economic model. In the background of this economic transformation, the major issues for target region to be addressed are concerted in how to balance the demand between endogenous developments and exogenous modulation and how to develop in a low carbon emission way which suited to local territoriality on the base of the local characteristics.

2.2 DRIVERS OF ECONOMIC TRANSFORMATION

In China "cities" are defined as "urban places" corresponding to local administrative and jurisdictional entities. There are three different administrative levels of cities in the Chinese urban system: municipalities (or province-level cities), prefecture-level cities and county-level cities. Small settlements with townships or lower administrative levels are not treated as "cities". The Chinese cities can be divided into 4 types which are Resource-based city, Industrial city, Integrated city and Tourism city.

The northeast part of China is characterized by the presence of vast flat land, abundant rivers and plentiful geology resources; the area affected by few geological disasters was fixed as the industrial base since PRC China was established. In this context, the majority of cities can be categorized into Resource-based city (e.g. Daqing, Fushun and Liaoyuan etc.) and Industrial city (e.g. Harbin, Shenyang and Changchun etc.). Compared with Integrated city and Tourism city, both the Resource-based city and Industrial city show higher values of energy consumption and carbon emission. With the sustainable development theory, the aim of the economic transformation which is leaded by national government is to guide the cities in northeast part of China to develop in a healthy way. The drivers of cities here should be transformed in an economic model that must take in high consideration two aspects: energy and environment. On the one hand, there is the heavy industry, the traditional pillar of industry, that, for improving the economic development and social construction in China, have worked at the cost of huge energy consumption, and that now has to be re-designed according with the scientific innovations and the energy limitation while the speed of economic development and economic aggregate are showing signs of recession; on the other hand, there is the environmental pollution which was caused by the long-term industrial production, and that becomes an even more serious issue during the winter time when the heating system was operated in the cities. Only the economic model is forced to transit from heavy industry pattern to intensive pattern to be presupposed; the low carbon development strategy should be explored as the one and only way for improving this area to be developed in a sustainable way.

The geographical location of northeast part of China is an important position and covers a great of area plentiful of natural and territory resources. So that, developing this clod climate area, especially the cities in this area, and promoting this area to develop in a low carbon sustainable way, not only shows important significance to the areas with similar climate, but plays also a significant role for improving social harmonious development and scientific development in the whole China.

2.3 A SIGNIFICANT RELATIONSHIP BETWEEN CARBON EMISSION AND THE CITY IN A COLD CLIMATE AREA

Addressing the issue of low carbon development in cities in northeast part of China, it should be made clear the importance of the link between the concept of sustainable development and the characteristics of the city. In recent years, with the urbanization process and social development in China, sustainability has guided the economic development transition from single extensive form to intensive form, and the social development prioritizing environmental targets and energy conservation to reduce pollution. Building on this background, in China the sustainable and low carbon city pattern is developing at high speed. According to the investigation by CSUC (Chinese Society for Urban Studies), China is the most active country in the world developing sustainable city: until February 2012, in China there were 233 cities that have set the development aim to become sustainable cities. This group encompasses 80.1% of all provincial level cities; among these 233 cities, there are 133 cities which are fixed "Ecological city" as the development aim, although the conception of "low carbon" hadn't been proposed until 2008, until 2010, there are 97 cities which fixed "Low Carbon city" as the development aim(Figure 1).

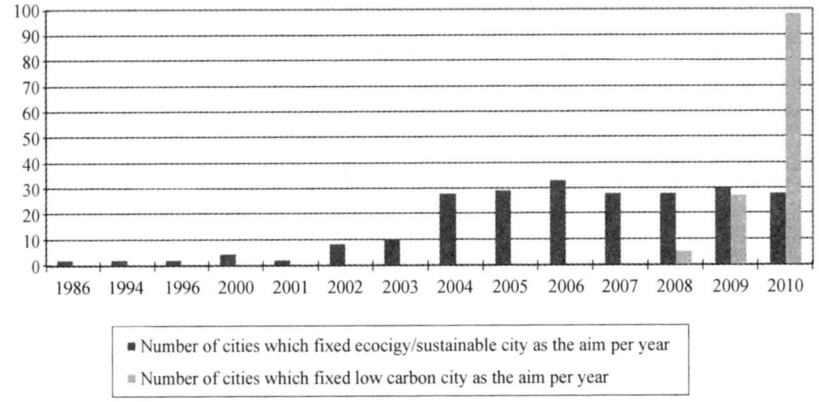

Figure 1 Number of cities in China fixed sustainable city/low carbon city as the development aim 1986-2010 (CSUS 2011)

Low carbon city defines cities where economic development has been decoupled from energy consumption and/or CO_2 emissions mainly in terms of building, transport and production . The carbon emissions in the cities are produced by industry, construction, transportation and infrastructure. We can measure the energy consumption by tracking carbon footprint and try to orient the city development in a sustainable way by perusing low carbon emission. The concept of "low carbon", introduced by the British government, is now fixed as the city development aim by many cities all over the world. "Low carbon" is the countermeasure that contrasts the environmental pollution since the industrial revolution. The aim of this statement is alleviating the energy crisis and climate change, especially for China which is in a high speed urbanization process, now that energy crisis and environmental pollution are becoming the key issues related to national interest and people's livelihood. Low carbon city is the inevitable choice for city development.

Ecological city, also named eco-city, has been referred to those cities that are self-reliant, minimize the demands on resources like energy and water and reduce waste. In the researches in China, the concept of "Green City" and "Sustainable City" are similar to Eco-city one. There is a renowned website named "Urban Ecology" that proposed the 10 principles for Eco-city as land-use high efficiency, transportation access by proximity, restoration of damaged urban environment; mixed housing; nutrition of social justice; local agriculture and urban greening project; support to recycling and innovative appropriate technologies; support to ecologically sound economic activities; promotion of voluntary simplicity and discouraging of excessive consumption of material goods; increase of the awareness of the local environment and bioregion.

Smart city was proposed as a new concept since 2009. With the new information technology, people can control the problems in the city and improve the efficiency of the city. This is a new conception which is aimed at improving the living environment in the city through coordinating the city operation by technology.

Actually, there are also other similar concepts which are often used such like Green City, Sustainable City, Knowledge City etc. The low carbon city model attributes the urban development to comparison of carbon emission in the urban area. Compared with the other methodologies, this method is more clear and utility. It's a good method for evaluation of the urban development. Thus the ten Eco-city principles summarized the key points in urban development process. Low carbon city strategy is more appreciate for politic decisions while the Smart City and Green City conceptions relate with more broad area.

Especially for cities in northeast part of China in cold climate, city development is influenced by local economic, available technology, traditional culture etc., in the extraordinary economic transformation period, to fulfil the aim of high performance, high efficiency, high benefit and low energy consumption, low pollution, low emission, the cities in cold climate area should fix the model of low carbon development, establish a research framework including urban design, urban planning, building design and evaluation. Therefore, low carbon city is necessarily the best choice for urban planning and building design development.

3 THE CHARACTERISTICS OF INDUSTRIAL BASE IN NORTHEAST PART OF CHINA IN ECONOMIC TRANSITION PERIOD

In the recent years, a serious of reform measures has been implemented in industrial base in northeast part of China, especially during "Eleven-five year plan" which is perused by central government; the economic structure has been promoted; the ability of independent innovation has improved and open up level has been raised a lot; these represent already a great achievement for cities that are suffering the resource depletion; the infrastructures had been improved, the key issues relates to people's living hood has been solving step by step, the social undertaking development has accelerated. This area is interested by a huge transformation. The economic transition plays an important role in the transformation process of cities and when we summarize the development track of industrial base in northeast part of China, the characteristics can be summarized into the following subsections.

3.1 THE INDUSTRY STRUCTURE DEVELOPMENT: FROM HIGH ENERGY CONSUMPTION TO HIGH ENERGY EFFICIENCY

As the heavy industry base of China, in the northeast area there were a lot of large state-owned enterprises, concerted in traditional industry. The industry development was established on the base of resource exploitation with serious environmental pollution. With the development of market economy, industrial layout was adjusted and the state government leads the transition from traditional industry to multi-industry, rising the energy efficiency and solving the problem of serious pollution.

Taking Harbin, city of the Heilongjiang province, as an example, the Dongli district was an area where many large scale state-owned enterprises were located. Actually, Dongli district was one of the administrative districts in Harbin, with an area of 131 $ksqm^2$ and 360,000 inhabitants.

Dongli district host 23 state-owned enterprises such as Harbin Electrical Machinery Plant, Harbin Boiler Plant, Harbin Steam Turbine Plant. In the original plan, this district was an industrial one and with its broad boulevards and wide sidewalks the transportation was good. There were 3 railways which length of 28 kilometres connected 16 large and medium-size enterprises. With the economic development, these state-owned enterprises were challenged by the market-oriented economy and needed to adjust their strategy structure. The industry layout would be arranged more flexible and would not depend on administrative interference to complete the production task. Under the economic transformation, these industrial structures should be optimized, the energy efficiency would be raised and the competitiveness should be improved. The economic model is changing from Intensive economical operation to Extensive model, from a high energy consumption and high pollution system to a high energy efficiency one.

In 2006, Dongli deflect was combined with Xiangfang district. Driven by industry adjustment, the urban space must be adjusted to adapt the urban development and supply better urban environment for inhabitants. For example, the Sandadongli Street located in this district, the original name of which was Daqing Street, is a typical street in this area. Along this street, the length of which is 1 500 meters, there are three big state-owned enterprises: Harbin Electrical Machinery Plant, Harbin Boiler Plant, Harbin Steam Turbine Plant(Figure 2 and Figure 3). Every day there are over ten thousand people working in these enterprises to accomplish the production target published by the national government. There are also residential zones connected with the factory district. The urban structure of this area is composed by industrial zones and residential zones. According to industry structure adjustment, the primary industry would take less part in economic structure, while the population would grow and the work choices would become more diversified. Following the development demand by itself, the new commercial districts and residential districts would break the original urban structure, the urban layout would change a lot.

Figure 2 Photos of industrial factory in Sandadongli Road

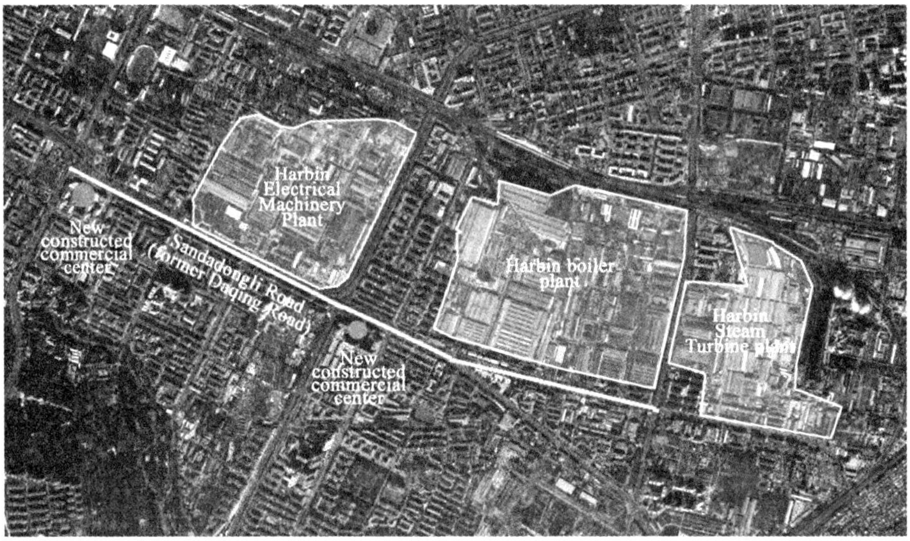

Figure 3 Sandadongli Road plan

3.2 SLOW GROWTH OF URBANIZATION PROCESS

The land is flat and fertile, the cities in northeast part of China were established because of the Middle-east railway had been built, while the urban construction is influenced by European architecture. Until 1943, the density of the railway network is 5 times that of the middle part of China. The railway construction accelerated urban development while, at the same time, its construction is a key factor for industry development in urban areas. Thus, influenced by the strategy of improving the railway infrastructure, especially after the reform policy was published, the urbanization process grows. But it does it slowly, in one hand because the regional development strategy lags, in the other hand, limited by local cold climate, the development of northeast part of China fell behind compared with other regions in China, in the same time.

3.3 THE ECONOMY IS UNDER DEVELOPMENT

In the past, the economic structure in northeast part of China was simple, so that the economic base was weak. As the social and cultural communication became more frequent, the former simple economic model did not fit with the social multiple development, and the economy developed slowly in northeast part of China, especially when compared with coastal areas in the south-eastern part of China, the economic situation in northeast part of China fell behind a lot.

The industry structure decide the economic form; economic form influence the urban development strategy and adopted technologies directly; so that if we want to fix low carbon development strategy for northeast part of China, we must make the research according to the local climate characteristics and fix the low carbon strategy to improve the local economic development.

4 THE EMERGING TREND OF URBAN DEVELOPMENT IN NORTHEAST PART OF CHINA

Now China is under the urbanization process, with the economic development and social improvement, the function of the city and inclusiveness became more and more important. The transportation improvement facilitates communication between different regions, in the same time, the communication exacerbates competition between different regions. To fix the development strategy which to fit the local situation, local characteristics must be taken into consideration. So that in the urban development, the local characteristics are the key point which should be thought about when urban planning and architecture design are made.

In allusion to the emerging trend of northeast part of China, we can summarize the following points:

- **Buildings' energy consumption is very high in urban areas**

According to previous research, carbon emissions in the building sector represent 39%, while carbon emission in transportation sector is 33% and carbon emission in industry sector is 28%. The city is the biggest carbon emission maker. China is developing country with high speed industrialization and urbanization—now China is facing the serious energy security issue because energy consumption in urban areas takes over 60%. The energy consumption per capita in urban area is 3 times that of people in rural area. In northeast part of China, this characteristic is more typical because in this region with four distinct seasons, winter time the temperature is quite low, the cost of building construction and building operation are higher. In one hand, with the improvement of living level, the requirement of living comfort level will rise, therefore energy consumption in building sector must rise as well. In the other hand, with urban area extending, the extending of building area causes the improvement of both building materials and energy consumption. At the same time, the background of economic transition will guide the high energy consumption by industry to decrease. So that in northeast part of China, building energy consumption will increase in the future.

• Sharp rise of the urbanization rate

Urbanization rate refers to the proportion of population in urban areas compared to the total population. Along with the society and science development, agricultural population is expected to decrease stepwise. Particularly for northeast part of China, since the land resource is plentiful, the land fits for agriculture. This area is the one of the key grain production base of China. In the past, cultivation was under development, and the agriculture population represented a big portion of the total population; within the development of agriculture automation and popularization of planting technology, there is a lot of people who is free out from agriculture works and enter the urban area to find a job opportunity. In northeast part of China, since the urbanization is accelerating, along with the industry structure adjustment, there are more economic growth points which attract more people living in urban areas. The result is the building area and urban area will rise sharply in the near future in China. This process was experienced in Europe decades ago. And the urbanization rate is not expected to stop increasing until the society reaches the balance of all industry distribution.

• Municipal system tends towards a higher wellbeing standard

According to the previous researches, the urban area is in direct proportion with per capita energy consumption. That means that when the urban area grows bigger, the energy consumption for each inhabitant would grow higher. That's because with the urban area extending, energy consumption for transportation grows; in the same time, with the municipal system tending toward increased standards of wellbeing, energy consumption for the municipal operation would rise higher, in connection with the promotion of the comfort level. In the past, energy consumption in urban area of China kept low level compared that of developed countries, that's in the worth of incomplete municipal environment and living environment. The development of "low carbon" city takes energy consumption or carbon emission as the measurement, aimed at changing the high energy consumption, high pollution which are caused by industry to low energy consumption, low pollution and high energy efficiency model. So that along with the development of city in cold climate, we should fix the urban planning and building design plan in accordance with municipal system tending towards wellbeing.

5 URBAN DEVELOPMENT STRATEGY IN NORTHEAST CHINA

In the background of economic transition, the development of cities in northeast of China located in the cold climate region must be characterized by low carbon emissions as the aim and guided by sustainable theory, we must propose the low carbon strategy specific for this region. We should use low carbon development model to transfer the environmental pressure from previous high energy consumption and high pollution; to control the city development from the point of view of urban planning, to adopt innovation technology and energy as measure, to improve the urban dynamic and to fulfill the final goal of an improved living standard.

Thus it can be seen, low carbon development strategy specific for northeast part of China is a complex evaluation system with multiple criteria to consider. If we think it from the point of view of climate, we should take two aspects in consideration: adverse impact caused by low temperature, wind and snow in winter; and the pollution caused by heating. The low carbon urban planning strategy could be summarized into five sectors: building, transportation, industry, spatial distribution and infrastructure. The low carbon development could be realized by three measures: decrease carbon source; carbon emission reduction; reinforce carbon capture. We should consider these five sectors and three aspects overall. As the previous analysis, low carbon development transfers the complex, multi-levels consumption to measurable and visualized issue. That means sum up the urban energy consumption from the view of energy. So we summarize the urban plan strategy into the following five sectors (Table 2):

Table 2 Low carbon urban planning in cold climate area

Category	Main factors	Target
Building	Building heating Building energy conservation measures Energy efficiency	Improving building energy utilization efficiency; Improving building environment comfort level; Reducing energy consumption per square meter.
Transportation	Transportation modes Public transportation system Private vehicle ownership Clean energy car	Improving urban transportation smooth; Improving transportation efficiency; Reducing transportation time; Reducing transportation energy consumption; Implementing public transportation; Implementing non-motor vehicle/pedestrian area.
Industry	Industry structure Adopting high-technology	Modifying industry structure in accordance with urban characteristics; Controlling high-pollution/ high energy consumption projects strictly; Developing high-tech industry, tourism etc.
Infrastructure	Utilizing clean energy Waste management system Water/electricity management system	Improving urban heating/water/ electricity system efficiency; using clean energy; Implementing water/waste recycle use; Reducing air/water/ territory pollution.
Spatial distribution	Urban density Land mix-use Green space	Controlling urban area, improving urban density; Improving land mix-use and efficiency; Improving green space in urban area and environment.

5.1 RECONSTRUCTING URBAN ORDER, INTEGRATING INDUSTRIAL AREAS

Since the cities in northeast part of China were built with the industry development, after the rapid development period and economic low tide period, the urban development strategy is needed to be proposed in the background of economic transition. There are many industrial cities in this area, except Harbin, Changchun and Shenyang, also included in Jinli, Anshan, Fushun, Qiqihar, Daqing, Benxi, Jixi, Fuxin, Mudanjiang, Hegang, Jiamusi, Shuangyashan etc. which were built step by step following the national development. For cities in different levels and characteristics, we have to emphasize the urban plan according to these differences. The original industry area occupied large land and became a complete social system which included kindergartens, hospitals, schools and residential buildings. These areas were located in the city center when the urban boundary extended.

We must readjust the land use according to the functions, reconstruct the urban order. For example, the case study of reuse of former location of Harbin Electricity Meter Factory which was founded in 1953. After the factory moved away, the original location had been developed to residential district and commercial center. The original factory was located in a main street named Xuefu road where there are several universities and middle school in surrounding area. Actually, because of the high density of universities and middle schools, the population density is much higher than average density. To balance the urban function, the reuse of this industry base was focused on residential building and commercial center which including restaurants, supermarket, cinema and shops etc. (Figure 4). After the regeneration of this area, the function of this urban area is more balanced and reasonable.

5.2 DEVELOPING SNOW & ICE AND INDUSTRY HERITAGE TOURISM TO GET MULTIPLIER EFFECT

In recent years, with the polarization of economic transition, there had been already some attempt of industry heritage renewal projects in China, some of them were successful. In cities in northeast part of China which are in the period of economic transition, we can use the economic transition as the development opportunity, develop snow & ice tourism, so that the cities can be changed from Resource-based city model to integrated city model or even to tourism city model to reach the aim of low carbon development. Specific for cities in northeast part of China, there are two typologies of tourism, one is industry heritage tourism on the base of industry base reuse; the other one is Snow & Ice tourism in cold climate area. Let's take Harbin as example, Harbin holds the International Snow and Ice Festival since 1985. In the International Snow & Ice Festival, except the traditional ice sculpture and snow sculpture exhibition, a serious effort towards diversification is made. The snow and ice sculptures are placed along the streets and squares(Figure 5). Harbin is trying to develop snow and ice culture which can attract more tourists and improve many other typologies of industries.

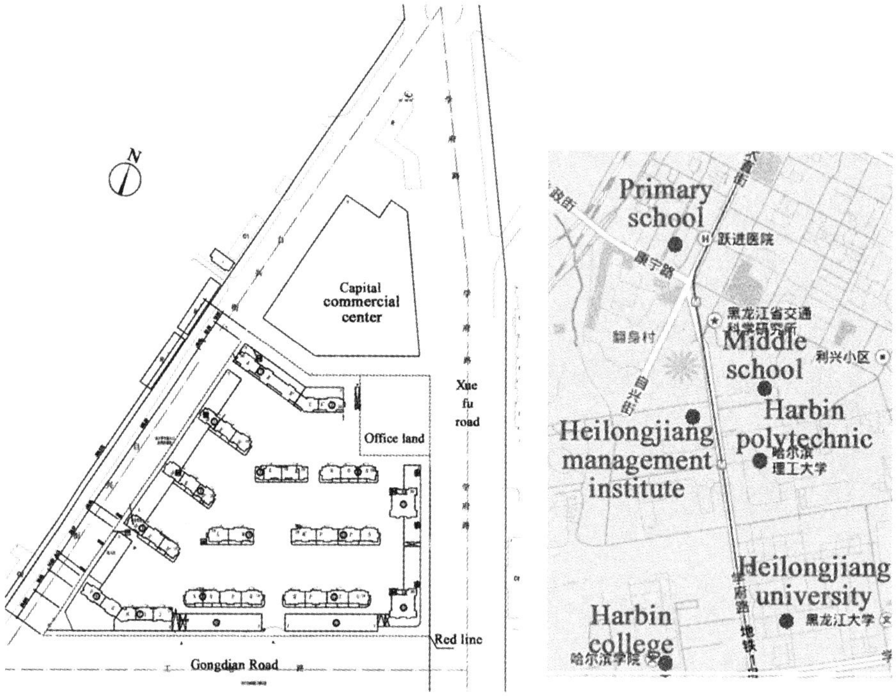

A Plan B Location analysis

Figure 4 Plan and location analysis of Harbin Electricity Meter Factory reuse project

In the meanwhile, Harbin can be set as the central point of snow and ice tourism and drive tourism in surrounding areas. We hope tourism development can raise the local economy effectively.

Figure 5 Snow & Ice utilization in tourism in Harbin

5.3 IMPLEMENT BUILDING CONSERVATION

Since low temperatures last long periods of time, the urban area of cities in cold climate should not be too big, something that can be achieved by increasing the urban density, with high-rise buildings to shorten transportation distances. Under the condition that their transportation plan is logical, cities in cold climate should improve the implementation of building conservation strategies, including the adoption of energy efficiency technology, innovative facade materials and evaluation systems specific for the local climate. Prof. Hongyuan Mei proposed a significant theory innovation for cold climate building design articulated through adaptation and moderation as the two design strategies for building design in cold climate urban area.

In the other hand, in China there is no comprehensive building evaluation system. In real practice, LEED (Leadership in Energy and Environmental Design) is used, but being a USA evaluation, there is no specific evaluation for the local economic and climate situation in northeast part of China. In the future research work, we are planning to set up a building evaluation system specific for this local situation.

There is a successful pilot in Harbin named Xishutingyuan residential district. In 2013, this project was awarded the national green building 3 stars level and it is the only one located in a cold climate urban area in China. In this project, zero energy consumption conception was adopted as the final aim for the demonstration building. Energy efficiency technologies are focused on three aspects: thermal performance of the envelope including the improvement of the facade, air tightness and selecting Low-E glass window; optimizing wind environment, which refers to the master plan of residential district by simulation; renewable energy utilization, which refers to GSHP (Ground Source Heat Pump). For example, the thickness of insulation layer of wall is 300 mm (Figure 6).

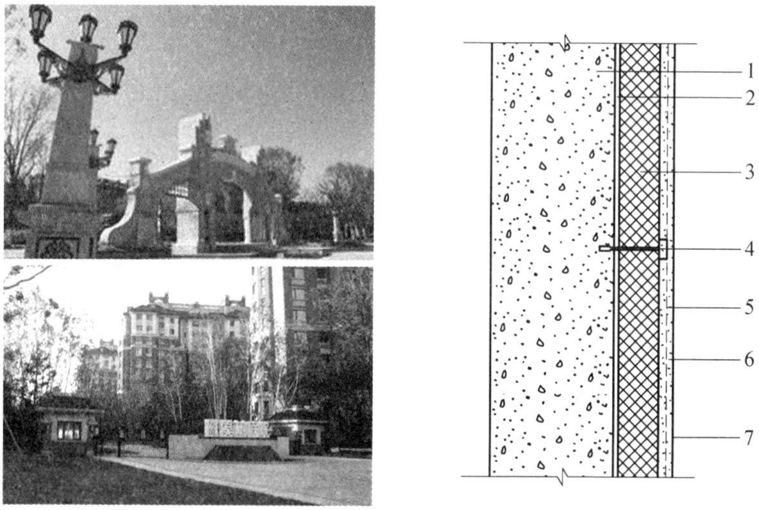

1—200mm ceramsite small-sized hollow block 2—adhesive 3—150mm HS-EPS module 4—bolts cross structure 5—Fiberglass Mesh 6—plasterer 7—finish

Figure 6 Photos and wall structure in Xishutingyuan

5.4 NETWORKED TRANSPORTATION GUIDING SYSTEM AND MUNICIPAL SYSTEM

In regards to transportation and municipal organization in the cities, cities aimed at low carbon urban development in cold climate conditions should be in harmony with a highly efficient transportation system, convenient public transportation and comfortable municipal system. Except rising urban density, land mixed-use is an effective measure for shortening transportation distances. In accordance with the demand of economic transition, we should fix reasonable urban special structure and then fix the networked transportation system through rail transportation system and public transportation, adding with private vehicle and convenient pedestrian paths. Especially, we should pay more attention on the utilization of underground spaces as a good method for saving energy consumption and improving comfort level. Utilization of underground space is an aspect which was neglected in China.

6 CONCLUSION

With the technology development and increased knowledge exchange, the issue of urban development becomes a problem related to interdisciplinary. The research about the city should be focused on the characteristics of the city and specific design strategy could be proposed. Urban and building development interact with each other and develop synchronously. In cold climate areas, the living environment is more severe in winter, energy consumption is much higher than in other regions, thus we should use the economic transition background as the opportunity to develop a low carbon city while improving the living environment.

ACKNOWLEDGEMENTS

This paper is supported by Fundamental Research Funds for the Central Universities: Energy management ssytem in residential building refurbishment project in cold climate area(HIT. NSRIF. 2014073) ;Planning annual project of Heilongjiang provincial philosophy social science research(14C018).

REFERENCE

[1]MEI H. Architecture in cold climate [M]. 1st ed. Beijing: China Architecture & Building Press, 2012.

[2]LONG W. Form and energy consumption future in low carbon city [J]. Architecture Science, 2010, 02: 13-18.

[3]CHEN F, ZHU D. Theoretical research on low-carbon city and empirical study of Shanghai [J]. Habitat International, 2013, 37:33-42.

[4]REGISTER R. Ecocities: rebuilding cities in balance with nature [M]. Gabriola Island : New Society Publishers, 2006.

[5]National Development and Reform Commission, Twelve-five years renewable plan, General Office of the State Council of PRC, 2011.

[6]LENG H. Habitability research for industry based city comparison [J]. Journal of Harbin Institute of Technology University, 2009, 09:13-17.

[7]GU Zhaolin. Climate change: carbon emission and low carbon urban development research[J]. Urban Planning Forum, 2009, 04:38-45.

[8]FU Y. Research of low carbon development route [J]. Impact of Science on Society, 2008, 04:5-10.

[9]PEI Z. Technology in passive low energy residential building in Xishutingyuan [C]// Proceeding of the 5th Youth Forum of China Green Building Council, Harbin, China, Harbin Institute of Technology Press, 2013: 82-85.

The Europe China Clean Energy Center

Roberto Pagani
Department of Architecture and Design, Polytechnic of Turin
roberto. pagani@ polito. it

ABSTRACT: Present paper illuminates the mechanism Europe-China Clean Energy Centre (EC2) and its effort in EURMUQI DEMO ZONE in Urumqi in China. As a demonstration zone for implementing clean energy in the city, the experience of EURMUQI DEMO ZONE could be spread to other cities in China and could be used to shape a more sustainable, environment-friendly and efficient energy sector in China.

Key Words: EC2, EURUMQI demo zone, clean energy

The Europe-China Clean Energy Centre (EC2) is a five-year cooperation project funded by the European Union. EC2 was founded in April 2010 by the European Commission, the National Energy Administration of China and the Ministry of Commerce of China, with the support of the Italian Ministry for the Environment, Land and Sea. It is managed by a consortium of nine partners—six European and three Chinese—led by Politecnico di Torino (Italy).

By supporting technological cooperation, providing policy advisory services, building capacity and raising awareness, the ultimate goal of the EC2 project is to promote the use of clean energy in China, thus contributing to Chinese Government efforts to shape a more sustainable, environment-friendly and efficient energy sector.

China is undergoing a massive urbanisation process. It is estimated that between 2010 and 2025, around 250 million citizens will move from rural areas into cities. The challenge is how to build or transform cities so as to avoid major problems related to energy, health, housing, waste and pollution.

Against this background, the Europe-China Clean Energy Centre (EC2) aims at promoting the wider introduction and adoption of innovative clean energy technologies through a combination of policy advice, research and technological analysis as well as through the identification of technological needs, potential projects, best practices and partner-

ship opportunities. In this perspective, EC2 has launched the EURUMQI DEMO ZONE initiative in China, focusing on energy and urban regeneration at local level.
The EURUMQI initiative (Energy & Urban Regeneration Using Methods Quality & Innovation) is supporting the city of Urumqi in low-carbon urban planning and circular economy via a clean energy demo zone, introducing successful EU experience and solutions. Replication in other cities in Western China is also envisaged.

Urumqi (Figure 1), the capital of Xinjiang Uyghur Autonomous Region in Northwest China, is the political, economic and cultural centre of Xinjiang province, a point of convergence between Asia and Europe and a gateway to Western China. The city comprises seven districts and has an administrative area of 14,000 km^2, including a constructed area of 339 km^2. Urumqi has 3.21 million inhabitants, 90% of whom live in the urban zone. It is a typical developing city, with an 11.2% growth in GDP recorded last year.

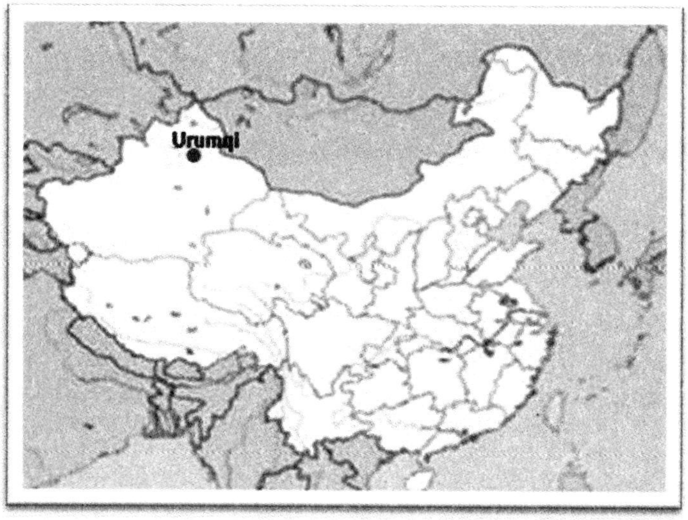

Figure 1 Location of city of Urumqi in the map of China

Urumqi adopted a 50% energy-saving standard for new public buildings from July 2005, and for new residential buildings from April 2003 (Figure 2). A 65% energy conservation standard was adopted in April 2009 for all new residential buildings. From the end of 2010, the energy-saving programme has been extended to 54.35 million m^2 of buildings.

Figure 2 Aerial view of Urumqi

The demo zone concept is a new area of exploration for EC2. Although the exact definition of a demo zone is still open to debate, according to the EC2 Strategic Plan a demo zone "represents the capability of focusing on a demonstration site, by using EC2 common methodological tools, which will unify joint actions and efforts. The demo zone(s) will be made available by the Chinese partners, while the EU partners will provide the professional and industrial expertise for the start-up of the demo zone design and technological solutions."

Demo zones have also been defined as physical zones with a cloud approach, comprising virtualisation followed by consolidation. This formulation highlights the strong methodological character of EC2's demo-zone related actions, through which Chinese cities selected as demo zones will be able to achieve efficiency and cost savings, as well as strategic solutions otherwise either out of reach or too technologically challenging. The demo zone concept covers activities related to energy efficiency, renewable energies, smart grids, low-carbon districts and smart community concepts, developed using an integrated approach. To test the concept, EC2 was assigned the challenge of drawing up a development strategy for the city of Urumqi, the capital of Xinjiang Uyghur Autonomous Region in western China. The city has embarked on an ambitious plan to reshape the structure of its industry, transform energy development, combat air pollution and reduce GHG emissions. Some 2,000 years ago, the city's location just off the main silk routes helped promote economic and cultural exchanges. Today, its competitive strength has given it a leading position among the five capital cities of western China, with the inevitable result that it is one of the most polluted cities in China and the world.

Urumqi needs to combine a new green economy with its recent development. To achieve this, it requires a comprehensive master plan with a fresh vision, new long-term targets and innovative concepts. The EC2 project therefore aims to: provide guidance to the city on the innovative, low-carbon planning of the proposed demo zone; set up a demo zone for clean energy development in Urumqi by introducing successful experiences in collaboration with EU cities and industries; and establish a replicable method and a dissemination mode for replication in other cities in western China.

In May 2012, EC2 and its consortium partners, the Energy Research Institute of the National Development and Reform Commission, Politecnico di Torino and Tsinghua University, as well as local partners such as Urumqi Development and Reform Commission and Xinjiang XinZi Engineer Consulting Co. Ltd., organised a Study Day event to launch the EURUMQI DEMO ZONE initiative in Urumqi (Figure 3). Local policy makers, administrators, public and private sector representatives and technology experts shared their vision for the future, explored what priority actions should be undertaken, and identified aspects to be reflected in policies for Urumqi (Figure 4). The 2030 timeframe was sufficiently long to allow for the introduction of broad transformations; and short enough for the results of their implementation to be seen.

Figure 3 Photo of Study Day Event for EURUMUQI DEMO ZONE

EC2 is now drafting a large-scale energy-oriented plan for the demo zone, with investments to be identified in the following main areas:

- Green buildings (refurbishment, new construction)

- Energy demand management in the industrial sector

- Innovative mobility solutions

- Green ecological industrial parks

- Circular economy industrial parks

- Resources utilisation development model

Figure 4 Participants of Study Day Event

The next step is to promote strong and effective public-private partnerships between the Urumqi government, public bodies and private companies. In addition, private companies will reflect European participation in demo zone development, enabling the full implementation of the EC2 collaboration scheme.

The demo zone is, primarily, a methodological challenge. Given its intrinsic complexity, the Urumqi demo zone challenge can only be met with a strong methodological background, creating a win-win strategy and advantages for both the local community and investors, and inspiration for other Chinese cities.

REFERENCES

[1]YU Wei, PAGANI R. City energy planning: an European strategy to reduce the GHG emission of the city[J]. Planners, 2009,25(3):62-68.

[2]MATTEOLI L. City futures: architettura design tecnologia per il futuro delle città[M]. MILANO: Hoepli, 2010.

[3] Bianciardi G, Cortassa C, Ellena G, et al. . Piano d'Azione per l'Energia Sostenibile di Torino / Tape turin action plan for energy [M]. Torino: Civicocentro stampa, 2010.